Jaro Gielens

Soft Electronics

Iconic Retro Designs from
the '60s, '70s, and '80s

gestalten

Table of Contents

4 Soft Electronics – The Liberation of Labor

10 Three Decades of Groundbreaking Design

16 Girmi Coffee Grinder Major *MC 14*
18 Kenwood Whisk *A 1050*
20 Philips Ladyshave *HP 2108*
22 Lady Braun Luftkissen-Trockenhaube *HLH 1*
26 Philips Special Ladyshave *HP 2116 FL*

30 The 1960s

38 Krups 80
40 AEG Kaffeeautomat *KF 1500*
42 Philips Coffee Maker *HD 5113*
44 Braun Coiffeur *HLD 3*
46 Braun Man-Styler *HLD 51*

50 The 1970s

58 Braun Multipress *MP 50*
59 Krups Biomaster
60 Lady Braun Super Hairstyling-Set *HLD 50*
64 AEG Foen 1000
66 Krups KM 40
67 Braun Aromatic *KSM 1*
68 Philips Ladyshave *HP 2111*
70 Lady Braun Cosmetic-Shaver

74 Braun

76 Braun 550
78 Braun Citromatic de luxe *MPZ 21*
80 Krups Coffina Super
81 Philips Coffee Grinder *HR 2109*
82 SHG Kaffee-Mahlwerk *MK 521*
84 Gillette Supermax Swivel
86 Krups Addigramm M
88 Braun Intercity
90 Krups Thermic Jet
92 Philips Die Schwebe Leise *HP 4628*
96 General Electric Power Turbo
98 Tefal Automatic Egg Boiler
100 General Electric The Looking Glass *IM-4*

102 Emide Elektro-Filterkaffeemühle *KM 11*
104 Bosch Mahlwerk-Kaffeemühle K4
106 Bosch Kaffeemühle K12
108 Krups Kaffee-Automat T8
110 SEB Cafetière Filtre
112 AEG Kaffeemühle *KM 101*
114 Krups Solitair
116 AEG Joghurtgerät *JG 101*
118 Tefal Super Mijoteuse
120 Moulinex Futura Electronic 4109
122 Moulinex Yogurt Maker
124 Moulinex Fruitpress

126 Moulinex

128 Moulinex Cafetière Espresso
130 Moulinex Electronic Egg Boiler
132 Krups Allround Styler
134 ITT Popcorn Party
136 Krups Duomat
138 Rowenta Waffelautomat Luxus *WA-02*
140 Bosch Elektro Messer *EM 1*
141 Krups Special Duo
142 AEG Foen Salon
144 ESGE Zauberstab *M 122*
146 Kenwood Blender *A 515*
148 SEB Mayonnaise-Minute
149 National Mayonnaise Maker *BH-906*
150 National Cooling Bag *ND-101*

152 National

154 National Curling Iron *EH 161*
156 National Speed Pot *NC-950*
158 National Sake Heater *NC-31*
160 National Cloth Dryer *ND-11*
162 National Ice Shaver *MF-U7*
164 Krups Suzette
166 Krups Party-Grill
168 Philips Rotating Grill *HD 4151*
170 Black & Decker Popcorn Center
172 General Electric The Looking Glass *IM-5*
174 Braun Protector *PG E 1200*
178 Moulinex Automatik Toaster

180 AEG Folienschweissgerät *FSG 102*
182 Calor Cafetière Expresso
184 Philips Water Cleaner *HR 7470*
186 Philips Mixer *HR 1187*
188 Rowenta Joghurt-Bereiter *KG-76*
189 Toshiba Yogurt Maker *TYM-100*
190 Krups Novodent Pulsar

192 The 1980s

200 Krups TS10 Aroma Super Luxe
202 Braun Multipractic Plus *UK 1*
204 Braun Softstyler *PG S 1000*
206 Braun Stabmixer Vario *MR 6*
208 Krups 3 Mix 4004
210 Moulinex La Pasta Machine
212 Toshiba Ice Crusher *KC-55A*
214 Philips Travel Iron *TI 6500*
215 Rowenta Fashion *DA-54*
216 Melitta Mr. Instant
218 Philips Air Cleaner *HR 4371*
220 Aromance Aroma Disc Player
222 Calor La Chocolatière
224 Rowenta Bimbo *KG-39*
226 Presto HotTopper
230 Kenwood Chefette de-Luxe *A 380*
232 Melitta Aroma Art
234 Philips Turbo Jet 1205 *HP 4125*
238 Philips Plus 400 *HR 2986*
240 Philips BOX 2 *HR 2010*

244 Philips

246 Philips Cafe Gourmet *HD 5560*
248 Philips Café Duo *HD 5171*
250 Braun Aromaster 10 Control S *KF 90*
251 Braun Aromaster *KF 43*
252 Krups Ovomat Trio

254 Index

255 Co-Editor's Acknowledgements

256 Imprint

Soft Electronics

The Liberation of Labor

More than kitchen companions, these appliances propelled society toward a different kind of future.

Design has a special ability to act as a marker of time. In many instances, it provides a starting point for further anthropological inquiry. In the case of design for the home, it showcases shifts in domestic life, revealing the gender roles within family structures across different decades.

Up until after the Second World War in the West, such product design was largely practical. Ornamentation was unimportant, brand styles were mute, and star designers were nonexistent. After 1945, new wartime materials and technologies challenged the status quo. Form and function became of equal importance, and machinery began to enter the household.

Despite feeling in some ways ever-present, the 1960s are as now culturally remote to us as the beginning of the 20th century was when the swinging decade began. It was a time when society was dreaming of different futures, a turbulent period of transformation. At home, women were still treated as housewives, but elsewhere, things were changing. As postwar economies began to stabilize, the average household income increased. Families, and in particular the women within them, were able to imagine a future that factored in day-to-day leisure time. Companies like Moulinex and Braun saw the opportunity to facilitate this, and began developing handy home appliances—defined throughout this book as "soft electronics"—which arrived with the promise of minimizing time and effort when it came to tasks and chores.

"Soft electronics are devices that were made for women," explains Jaro Gielens, who coined the term and whose collection of products forms the basis of this book. "Rather than products designed for men, like audio/video equipment and tools, which I would describe as 'hard electronics.'" The design of these devices has what Gielens describes as "a different baseline." This baseline dictated that they needed to be easy to use, look good in a home domestic setting, and, to ensure they didn't create more work in place of the work they'd alleviated, they had to be easy to clean.

Plastics prevailed and, thanks to marketing departments becoming more involved in the development of new products, top brands would use colors to distinguish themselves from their competitors. Having multiple colors of the same product was considered state of the art. French company Moulinex did many two-color versions of the same device, whereas fellow French company SEB and German company Rowenta rarely had colored

Companies began developing handy home appliances which arrived with the promise of minimizing time and effort when it came to tasks and chores.

versions at all. Dutch firm Philips only changed colors when updating a device as part of a new product cycle.

Given that soft electronics were often bought as presents, packaging was also of the utmost importance. The product needed to look enticing on the shelf and special enough to be bought as a gift. Many of these devices were brand new to the market—so the branding on the box needed to be self-explanatory and sales-oriented. "It was the first time that consumer electronics were presented with packaging," explains Gielens. "Previously, this kind of branding was only used for convenience goods like food and cleaning products."

As more and more devices came to market, electrical stores would hold tuition sessions in order to get housewives—their users—up to speed on the latest technologies. The pages of popular women's magazines were filled with articles and reviews based solely on household appliances, and television advertisements would feature women as a way of selling the product from the same stereotypically gendered perspective.

It would be difficult, and of course inaccurate, to credit the liberation of women to a few kitchen appliances. Yet, the fact that attention was now being paid to the way they did things around the home and how their experience could be made better had a huge impact.

In his 2019 book, *Evolving Households*, Jeremy Greenwood argues that technological progress has had as significant an effect on households as it has had on industry. Looking at trends—such as the rise of married women in the workplace, the declining number of marriages, and changes in fertility rates—he observed a dramatic transformation in everyday life, and deduced that it was thanks to technological advances inside the home taking place within the concurrent time period.

"In the 1800s, the mother in most American households worked at home surrounded by six children," he writes in the introduction. "Housework was laborious in a world without running water, central heating, and electricity. The Second Industrial Revolution introduced electricity and labor-saving household appliances. Additionally, the value of physical strength declined on the labor market as machinery took over strenuous tasks. It is not a stretch to say that these developments liberated married women from the home."

With Greenwood's points in mind, to be a futuristic household became the ultimate luxury, and before long, everything from boiling an egg to making mayonnaise came equipped with a small plastic machine. Looking back, many of the products seem somewhat useless. Even by today's standards, using a machine to cut potatoes into squares when it can be done as efficiently by hand is arguably counterintuitive. A machine takes time to clean, whereas hands and a knife can be washed and dried with ease.

Funnily enough, many of these devices that were created with the promise of bettering a woman's life were actually designed by men, who, at the time, were not so au fait with the kitchen. "That is probably why some designs were not as successful as others," adds Gielens, "because some designers didn't understand what women needed or how they would use a device ideally."

And while the egg boilers, mayonnaise makers, and potato cubers may not have stood the test of time, products by one designer in particular, Dieter Rams, seem to have correctly catered to the needs of the housewife, with many of the products he created for the German firm Braun still part of the brand's catalog today.

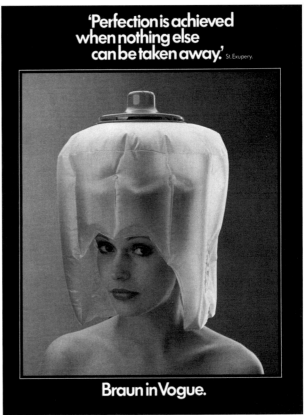

Rams joined Braun in the 1950s as an architect and interior designer and eventually became the company's chief design officer, a position he held until 1995. During his 40 year tenure, he developed a vision that would go on to inspire generations worth of designers across a spectrum of disciplines. His ten principles for good design, immortalized in the 1970s, spoke of sustainability before it became a tired-out buzzword, and provided a framework in which he believed other designers should work to produce meaningful objects. The ten principles argued that good design should be: innovative, useful, aesthetic, understandable, unobtrusive, honest, long lasting, environmentally friendly, and lastly, minimal. Ultimately, he believed obsolescence within design was wrong.

Despite this, by the 1990s, many small appliance companies saw Rams's principles as outdated. Quality lessened, manufacturing moved almost completely to Asia, and planned obsolescence became key to product design. Products were being sold to consumers by companies who knew that they had a short lifespan. While 30 years ago, you may have been reluctant to pay for a product that you knew wouldn't last—these days, it's hard to escape.

The figures are astounding. In 2019, more than 50 million tonnes of electronic waste was generated globally, with only around 20 percent of it officially recycled. As consumers have woken up to the issues associated with this throwaway electronic culture, companies are beginning to be held accountable. In 2020, Apple, whose former Chief Design Officer Jony Ive, ironically credits Rams's principles as an inspiration, agreed to pay up to 500 million dollars in settlements related to allegations that software updates caused older iPhones—such as the iPhone 6, 6s Plus, 7, and 7 Plus—to slow down, prompting customers to purchase more recent models.

Big changes from big companies are undeniably needed, but as consumers, we too have the ability to make informed decisions about what we buy and how we care for it. With the climate emergency posing an urgent threat to the world as we know it, it will be interesting to see how—just as they changed the culture of the '60s, '70s, and '80s—electrical products for the household, and the companies who create them, can help us to adapt our habits too.

Three Decades of Groundbreaking Design

1961
The legendary industrial designer Dieter Rams is appointed head of Braun's newly established design department. Under his command, the German company would soon blaze a trail in modern electronics design.

1962
American company Sunbeam develops the "spray mist" function for its irons. Alongside its preexisting "steam" and "dry" functions, it marks a vital new chapter in ironing technology.

SPRAY AS YOU IRON— AT ANY SETTING

1960
U.K. brand Remington introduces the first rechargeable battery-powered electric razor. Convenient (it eliminated cords) and cost effective, this was a vital step in the evolution of shaving devices.

1961
Italian brand Faema releases the first pump-driven espresso machine. Instead of physical force, it uses a motorized pump to push the water through the coffee. This design becomes the template for future espresso machines.

1963
U.S. brand General Electric introduces the first self-cleaning electric oven, followed by the first electronic oven control in 1967. These key developments preempt the use of microprocessors in household appliances.

1963
U.S. brand Hoover introduces the first clean-air upright vacuum cleaner to the market, confirming the power of hard plastic. The design inspires countless spin-offs from competing manufacturers.

1964
Fondue is marketed in America for the first time at the New York World Fair. Within a decade, electric fondue makers will emerge as the kitchen gadget *du jour*, the ultimate 1970s party piece.

1964
Braun introduces the HLD 2 blow-dryer. Minimal, with a unique, boxy shape, the plastic design is one of many lightweight, affordable options to emerge during the '60s, rendering the blow-dryer a household staple.

1963
Dutch brand Philips creates the first compact cassette audio player, signaling the start of a home entertainment revolution. VCR players, video games, and the personal computer will emerge within 11 years.

1967
Braun releases its iconic Aromatic coffee grinder. A sleek, pared-back, and fool-proof design, it epitomizes the brand's output at the time and has since become a coveted collector's item.

1971
French inventor Pierre Verdon debuts the first compact domestic food processor, Le Magi-Mix. Tasks like chopping, grating, and slicing can now be done quickly by a single machine. Processors will soon become a kitchen must-have.

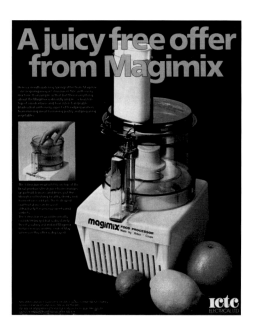

1970
Four years after the release of its first electric women's razor, Philips introduces its popular HP 2108 model. By the mid-1970s, it is producing 1.5 million women's shaving devices annually. Braun would release the rival Lady Braun Cosmetic-Shaver in 1972.

1970
Refrigerators are now a standard and integral part of the modern home, reflecting the extent households have embraced time-saving home appliances, including washing machines.

1971
The microprocessor is invented by the U.S. company Intel. The device accepts binary data as input, processes the data, and delivers output based on memorized instructions. It will revolutionize household appliances.

1972
The Crock-Pot slow cooker is released by The Rival Company in Missouri, USA. A driving force in the popularity of slow-cooked meals, it sparks the arrival of countless competing appliances.

1972
Eighteen years after the invention of The Wigomat, the first electric drip coffee maker, Sunbeam introduces Mr. Coffee, a percolator with an automatic-drip process and pre-programmed cut-off control to lessen the risk of over-brewing. Such machines become *de rigueur* thereafter.

1977
German appliances brand BSH launches the first under-the-counter dishwasher, the ultimate, space-efficient time-saver.

1978
French company Moulinex creates the innovative Automatik Toaster. Picking up on the latest toaster trends, it features wider, multiple slots, offering greater compatibility for different types of bread.

1976
American manufacturer Singer unveils the world's first electronic sewing machine, the Athena 2000. The pioneering device offers a wide variety of stitches at the touch of a button, changing sewing forever.

1975
Five decades after the domestic freezer was invented by General Electric, the appliance has become affordable enough for popular use. One in three U.K. households now own one, and the frozen food fad is underway.

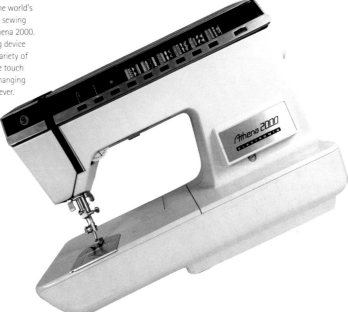

Three Decades of Groundbreaking Design | 13

1981
Moulinex releases its miraculously multi-tasking slow cooker, Le Cuitout, which can cook, stew, fry, steam, and boil. The most innovative slow cooker yet, it sets new standards for the device.

1982
Braun releases the Softstyler, a blow-dryer with a difference. Harnessing the '80s maximalism, the design comes with a bulbous detachable diffuser to achieve the ultimate big hairdo.

1980
The microwave, patented by Percy L. Spencer in 1945, is now affordable and widely available in countertop form, heralding a new mode of cooking. Fast, microwavable meals will prove an '80s favorite.

1981
Braun introduces the MR 6, a powerful, easy-to-grip hand blender that is better at puréeing than any rival. It is a key development in hand blender design, which gains traction in the '80s.

1983
Philips develops its influential Café Duo, a compact coffee maker that produces one to two cups at a time. It reflects a new type of consumer—the busy, single business professional.

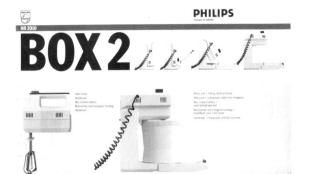

1984
Apple introduces the Macintosh 128K, the world's first mass-marketed personal computer—it marks a new age for home and office technology. Steve Jobs demonstrates its revolutionary potential in the first of his famous Mac keynote speeches.

1985
German brand Rowenta introduces its nifty Fashion travel iron, replete with a fold-down handle: one of many emerging designs targeting a new generation of business travelers.

1983
As part of the BOX series, Philips releases the BOX 2, a food mixer that can transform from a stand mixer to a hand one in seconds. Home appliances are more multifunctional than ever before.

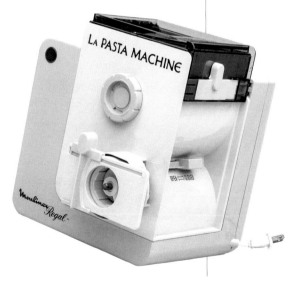

1985
U.S. businessman Joe Pedott patents The Clapper, a sound-activated switch that allows users to activate appliances by clapping. Initially viewed as a novelty item, it proved the original precursor to smart tech.

1983
Moulinex debuts its complex pasta maker La Pasta Machine, which can mix, knead, and shape pasta dough—another example of the increasing demand for fanciful household appliances.

Three Decades of Groundbreaking Design | 15

Coffee Grinder Major

Girmi | Model No. MC 14
Italy, 1965

This squat orange mushroom is the enduring coffee grinder by Girmi. The purity of its design has an almost brutalist feel to it, with its sturdy, top-heavy proportions. The power cable wraps around the base and rather than being a nuisance—as is often the case with kitchen appliances—becomes a design feature in itself. Unfortunately, the parts don't align perfectly, but that's just part of the charm, as is the softer plastic that it's made of. In 2016, the grinder was on the cover of the book that accompanied the *Cucina & Ultracorpi* kitchen design exhibition at the VIII Triennale Design Museum in Milan.

Coffee Grinder Major | 17

Whisk

Kenwood | Model No. A 1050
United Kingdom, 1973

Named simply Whisk, Kenwood's early 1970s mixer is minimal in both its name and its form. This humble, English-made hand mixer is modest in size, is battery powered, and comes with an orange plastic mixing wand with two overlapping circles at its tip. The design, typical of Kenwood at the time, is by British industrial designer Sir Kenneth Grange, who would go on to design Kodak cameras, Parker pens, and Imperial typewriters, as well as the iconic 1997 LTI TX1 London taxi cab, which can still be seen zipping through the capital today.

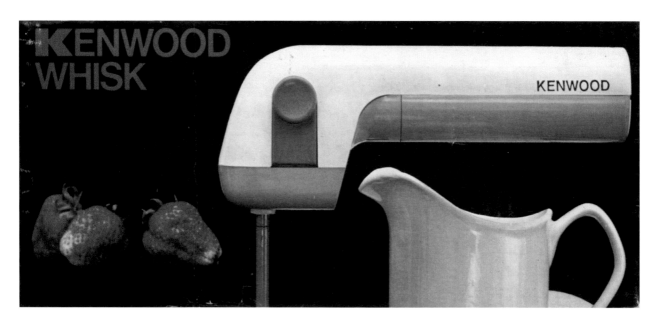

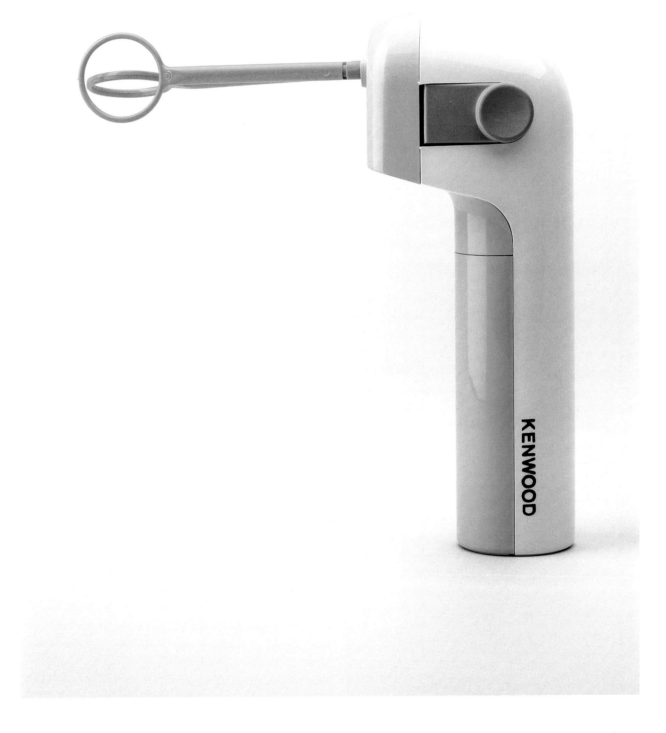

Ladyshave

Philips | Model No. HP 2108
Austria, 1970

Six years before Braun released its first electric women's shaver under the now-defunct sub-brand Lady Braun, Philips began selling its own version. Originally released in 1966 and designed by Willem Janssens, the Ladyshave model pictured here was rereleased alongside the Ladyshave Special HP 2116 FL (page 26), which came in an extravagant gift box. The Ladyshave functioned using foil blades, and was produced in a purpose-built factory in Klagenfurt, Austria. By the mid-1970s, the factory was producing more than 1.5 million devices annually.

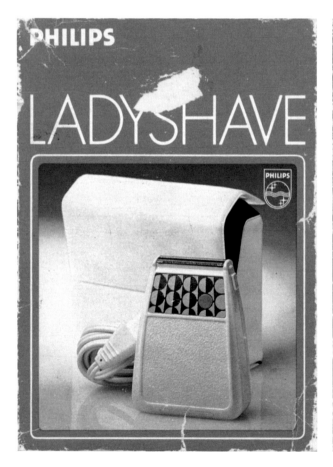
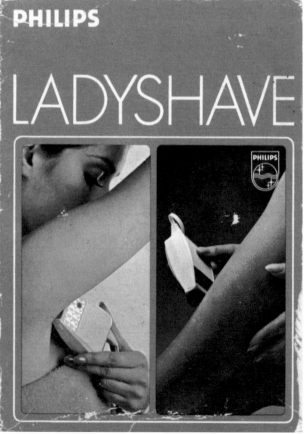

Lady Braun Luftkissen-Trockenhaube

Braun | Model No. HLH 1
Germany, 1971

Although they're still used today, bonnet hair dryers speak to a bygone era of beauty regimens. And this product—with its packaging and a name that translates as "air pillow"—is no exception. The Luftkissen, released by the now defunct women's sub-brand Lady Braun and designed by Jürgen Greubel, is among the most iconic at-home bonnet designs, and would go on to be imitated closely by several other brands. Its futuristic design allowed hands-free hair drying, and it was operated through an on-off switch in the power cable.

Lady Braun Luftkissen-Trockenhaube

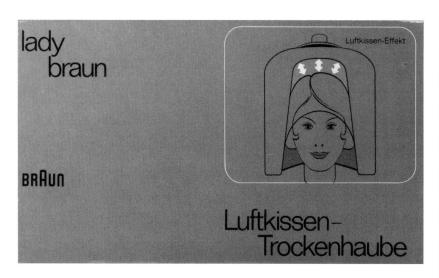

Braun only used models on the packaging when the product needed to be shown in use, e.g., for personal care products, to make their function more understandable.

Luftkissen– Trockenhaube

Special Ladyshave

Philips | Model No. HP 2116 FL
Austria, 1973

This Günter Hauf design was a higher quality, supersized version of Philips's popular Ladyshave electric razor, and offered the best alternative on the market to Lady Braun's Cosmetic-Shaver. The case—which opens up like an oversized compact mirror and is much larger than the product it conceals—is a status symbol, revealing a top-of-the-line shaver suited for spacious bathrooms with the capacity to accommodate it. Products like this were often bought as gifts, and that's where the case design becomes a helpful marketing tool. With space for the spiral cable to encircle the razor, the case is just one of many different shapes and sizes that cosmetic shaver designs adopted over the years.

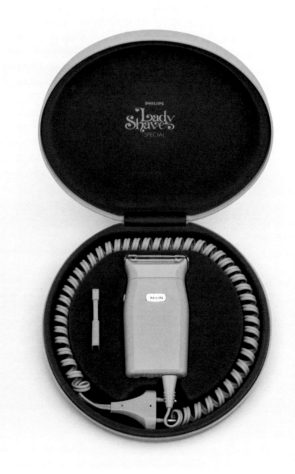

The case was a status symbol and the shaver was often bought as a gift. Packaging photography by Christopher Joyce and packaging design by Henk Jan Drenthen.

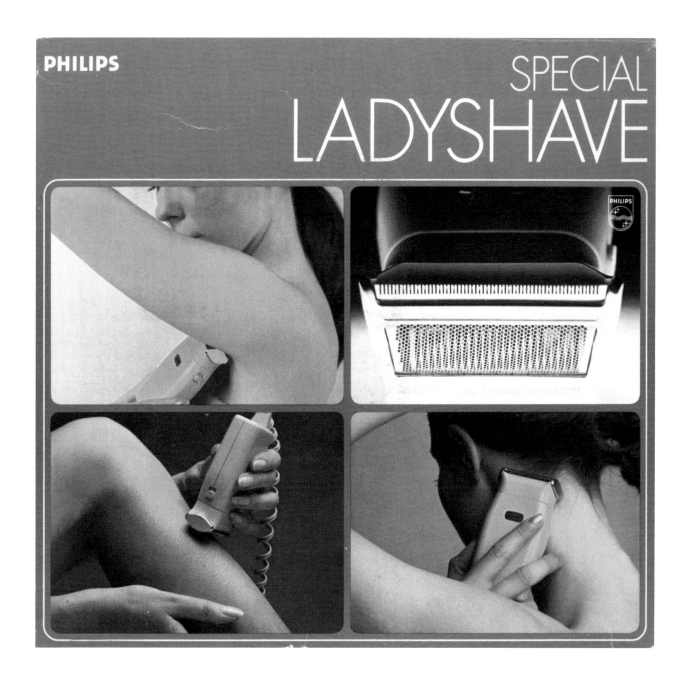

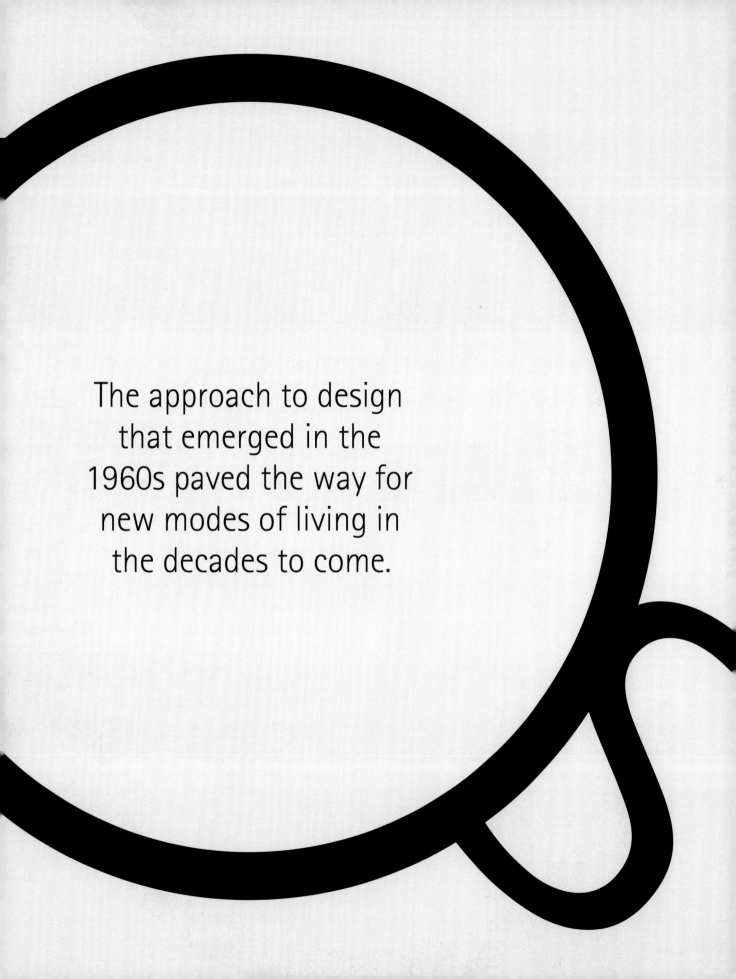

The approach to design that emerged in the 1960s paved the way for new modes of living in the decades to come.

The 1961 British advertisement for Formica laminated plastic work surfaces. General Electric magazine advertisement for new products.

In the 1960s, the world was rapidly changing: political beliefs were in turmoil, countercultures boomed, and experimentation in life and art reigned. Liberated from tradition, design entered a new era—one defined by technological advances, the democratization of consumer products, and an increasingly light-hearted approach to interior design. A widening generation gap meant younger and older people were living what looked to be completely different lives. Second-wave feminism arrived to further the fight for equality, with a broadened view that included issues like reproductive rights, family, and the workplace. Yet, despite the growing influence of feminism, many women remained trapped by tradition in the home.

A research project beginning in 1967, undertaken by Swedish-born architect and industrial designer Sigrun Bülow-Hübe, declared that Canada's five million housewives were working in inadequately designed kitchens. Bülow-Hübe's research surveyed women at work in these cooking spaces; mapping how far they walked, the height and space provided by counters and cupboards, and the back-and-forth required to cook a family meal. The study illustrated that the '60s kitchen was both a technologically complex and inherently gendered space. Bülow-Hübe concluded that L-shaped kitchens provided the most efficient workspace for women, a design blueprint that would remain popular in the West for decades to come.

In addition to shifting floor plans in the kitchen, changes were made in other domestic areas to make life easier for increasingly busy homemakers. Rather than considering a redistribution of household responsibilities, designers focused on creating products for the '60s housewife—developing appliances that reduced cooking and preparation time in materials that were easy to maintain and clean: among them, laminate, linoleum, plastic, and Bakelite. New brands, including Philips, AEG, Bosch, Sunbeam, and Moulinex, also entered the home-appliance market, bringing with them further innovation. The decade saw the introduction of the hot iron, vacuum cleaner, and the blow-dryer—which was notable for being a somewhat dangerous device that was difficult to use in its early iterations.

The abundance of cultural change taking place across society was reflected in the period's design, with homes seen as status symbols and ideas of the ever-present future. Outside, newly built houses came equipped with driveways to make room for multiple vehicles, another indication of wealth, and inside the obsession with psychedelia saw shag carpets paired with wallpapered surfaces and modernist Danish furniture.

New technology and the rise of convenience food caused a decline in traditional cooking. This was the era of classic science fiction shows such as *Star Trek* and *The Jetsons*, a time when home appliances were looking to the future also, with the introduction of automatic dishwashers, the microwave, multiple-burner cooktops, and fridges with ice and water dispensers.

British advertisement for Perspex bathtubs—warm, beautiful to the touch, and easy to clean.

Fridges, which became common in households during the 1940s, proved to be one of the most revolutionary appliances of the '60s. Many came with recipe books designed to assist housewives in the culinary art of "cold cookery." Refrigerated meals would often involve cooling or setting time, freeing up a couple of hours for women to get on with something else, which, given that it took up, on average, seven to nine hours of their lives per day, would probably be cleaning.

When it came to the design of small household appliances, German firm Braun led the way. The company's head designer, Dieter Rams, defined the 1960s appliance aesthetic, and in turn, preempted the look of home electronics for decades to follow. Rams's illustrious career made a lasting impact on industrial design; his "less, but better" philosophy and his 10 principles of good design focused on sustainability and functionality. The fact that many Braun products—like the Citruspress, first released in 1965, which has continued to evolve as a line with iterations available for sale in the 2020s—are still sold today, is a testament to the influence of Rams's ethos.

While the traditional view of family life was still very much apparent during this period, people began leaving the nest at a younger age—daring to dream of lives different than their parents'. This generation gap was immortalized by The Beatles in "She's Leaving Home," a song which was based on the true story of a 17-year-old who ran away from her family's comfortable North London home in February of 1967. Given that a large proportion of teenage income during this period was disposable, the booming youth culture of the '60s was well-funded. Music, art, drugs, antiestablishmentism, protest, and the sexual revolution are intertwined with the lasting memory of the decade.

For those families who were still together, this period also brought about a social change when it came to mealtimes. Rather than having a main meal in the middle of the day, families would eat a bigger meal in the early evening, and it would be eaten

The floor plans of kitchens changed to make them a more efficient work space for women. Changes were also made in other domestic areas to make life easier for increasingly busy homemakers.

New technology and the rise of convenience food caused a decline in traditional cooking.

together, around the dinner table. The chairs that they sat on were colorful, often made of plastic, and designed by the coveted names of the period: Pierre Paulin, David Hicks, Verner Panton, and Terence Conran.

Conran, based in London, had been designing under his own practice since 1956—his most notable projects at that time being the teak Summa furniture range and a shop interior for Mary Quant, the designer who popularized the miniskirt and hot pants during the Swinging Sixties. In 1964, the first Habitat shop, which focused on contemporary homeware and furniture, opened in London's Chelsea. Thereafter, Habitat grew into a large chain and is often credited with being the first retailer to bring such designs to a mass audience.

Similar franchises were appearing elsewhere too, particularly among grocery stores. The rise of the supermarket took place in the '60s, bringing with it the beginning of the self-service economy. Relieved from some of their household duties thanks to new appliances, women would head to the supermarket to pick out items for themselves. Prior to this, it was common for food stores, butchers, and fresh-fish retailers to fetch people what they needed. The supermarket offered a more flexible way of shopping, and in turn, a more imaginative way of eating.

One could argue the '60s was a decade of opposites, largely because of the extreme highs and lows it experienced in both politics and popular culture, as well as the polarizing opinions that rippled through society. But in the home, it offered a glimpse of a different kind of future, one in which women were released from their kitchen duties and able to live a life outside the realm of domesticity. While that version of the future would take a few decades to arrive, the '60s certainly saw its foundations put in place.

Krups 80

Krups | Model No. Krups 80
Germany, 1971

This chunky electric pocket shaver came with its own handsome carrying cassette, opening up to reveal the razor and a small mirror as well. It's an earlier example of a pocket shaver—modern in its design but still humble. The Krups 80 is the German manufacturer's answer to Braun's Sixtant 6006, which was similar in form and also came with a storage cassette featuring an enclosed mirror. Nevertheless, the Krups 80's finishing is solid and high quality. The brand would later release orange versions under the names Flexonic II and Flexonic Junior.

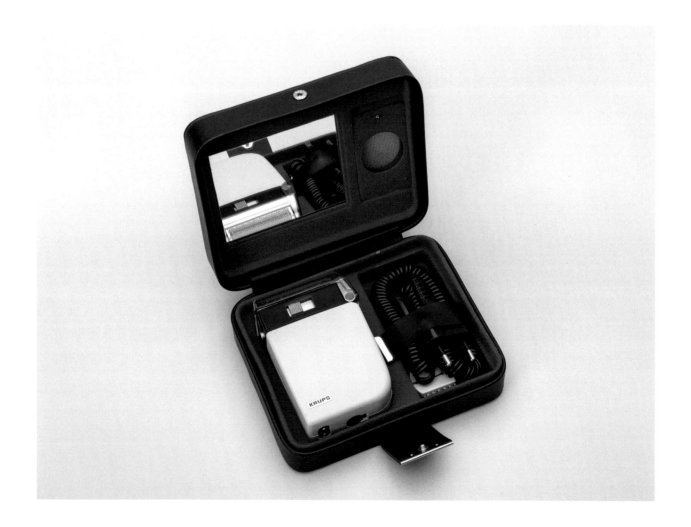

Kaffeeautomat

AEG | Model No. KF 1500
Germany, 1970

The world's first patented electronic filter-coffee machine was the Wigomat, produced in Germany in 1954. The model shown here by AEG is an evolution of this original, its design combining classic 1960s elements with chunky '70s plastic, hinting at the fact it was released at the turn of the decade. The filter holder, carafe, and water container are all made of the same durable Jena glass, with thick, hard-wearing plastic components. All of these elements contribute to the machine's overall heft, but that, too, is a benefit: it can brew 10–12 cups.

Kaffeeautomat | 41

Coffee Maker

Philips | Model No. HD 5113
Netherlands, 1973

The rectangular base and boxy form of this Hans Jülkenbeck design is typical of early Philips products and was one of their first modern plastic coffee machines. Unlike their French and German competitors, the Dutch household brand focused on creating compact coffee pots during this period. After releasing the HD 5113 in 1973, Philips went on to re-release it with slight color tweaks, different materials, and new design features (such as a retractable water outlet and a power switch).

Coiffeur

Braun | Model No. HLD 3
Germany, 1972

Available in both black and white, this blow-dryer by Reinhold Weiss is an updated version of the HLD 2, released in 1964, also designed by Weiss. It takes cues from Dieter Rams's HLD 4 (1970), which more resembled a brick than a hair product. The 1972 iteration, shown here, unifies Weiss's previous designs with Rams's into a more ergonomic shape, but keeps Rams's circular on-off switch instead of a rectangular one. Weiss worked with Braun throughout the 1960s and 1970s, designing, among other items, toasters, coffee grinders, and kettles.

Coiffeur | 45

Man-Styler

Braun | Model No. HLD 51
Germany, 1972

As evidenced by its packaging, this product was geared towards styling luscious 1970s hairstyles. The HLD 51 was the same product as the HLD 5, the "Lady Braun Hairstyling-Set," which was released in the same year, save for a few differences—most notably that this one was targeted toward men. This meant that, unlike the female counterpart, this product came in black rather than orange. Aside from this, the products, designed by Heinz Ulrich Haase, Reinhold Weiss, and Jürgen Gruebel contain identical elements, and were just sold under a different name.

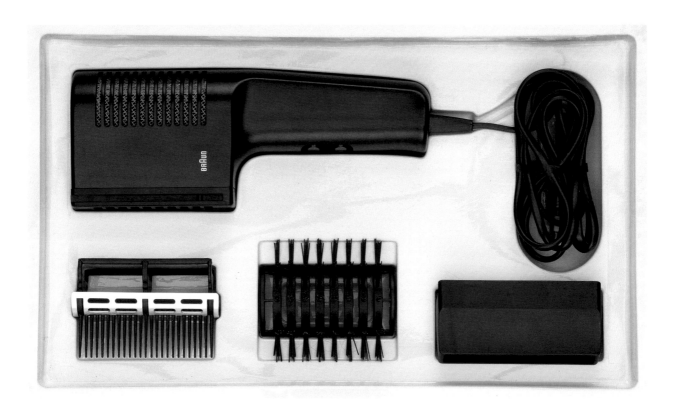

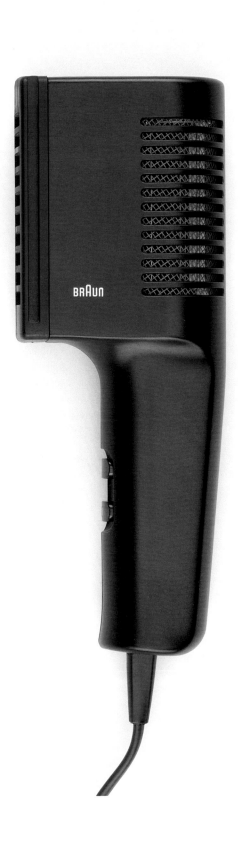

The Man-Styler was one of the rare black products and was intended for a male target group.

Braun Man-Styler

BRAUN

Man-Styler

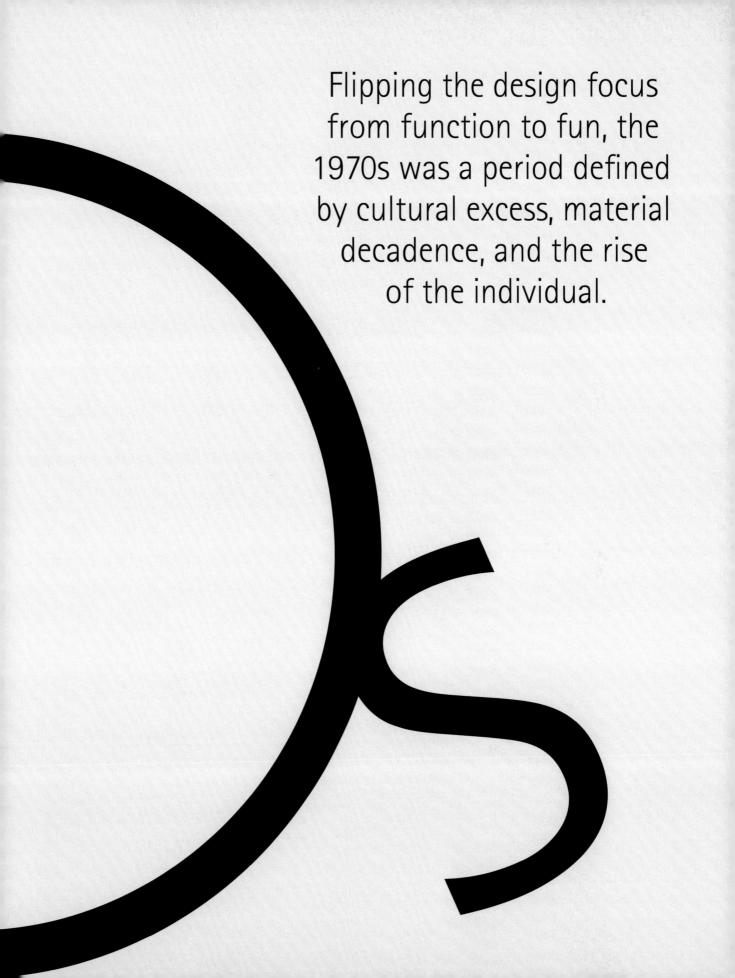

Flipping the design focus from function to fun, the 1970s was a period defined by cultural excess, material decadence, and the rise of the individual.

People were turning inwards and finding comfort for themselves in luxuries. It was a time of interior decadence, waterbeds, disco, bell-bottomed pants, and sexual liberation.

n 1976, American writer Tom Wolfe coined the phrase "the 'Me' decade" in a now-famous essay published in *New York Magazine*. Wolfe, born in 1930, was referring to the societal shift that he had observed in the West between the 1960s and 1970s, describing the '70s as being more concerned with individual pursuits than the political and social issues that drove previous decades.

Despite the '70s being revolutionary in many ways—with both feminist and environmental movements gaining further traction—political conservatism was still very apparent. The strong period of economic growth that took place postwar ground to an abrupt halt at the end of the 1960s, resulting in a recession across much of the West between 1973 and 1975. The combination of high unemployment rates, low growth, and inflation meant that the prices of goods, including food, were increasing faster than the average household wage. Many women were forced out of the home and into paid work to make up for the loss in income.

Despite the financial uncertainties facing many households, society was moving away from restraint and towards excess. As Wolfe observed, people were turning inwards and finding comfort for themselves in luxuries. Homeowners were modernizing their surroundings along with their appliances, and spaces were made party-ready. Conversation pits saw sofas sunk into the ground ready for intimate encounters, garish patterns and wallpaper abounded, as did a nature-inspired color palette spanning reds, oranges, browns, yellows, and greens. It was a time of interior decadence, waterbeds, and disco, of flared trousers and sexual liberation.

At the beginning of the decade, feminism had very much arrived, particularly among the younger generations, and the women's rights movement made significant and successful challenges to sexism. Popular books such as *Sexual Politics* (1970) by Kate Millett and *Sisterhood is Powerful* (1970) by Robin Morgan brought the concept of feminism to the masses. Pioneering artists questioned why their female peers remained unacknowledged. Judy Chicago's sculpture *The Dinner Party* debuted at the San Francisco Museum of Modern Art in 1979, with a triangular table made up of 39 place settings each dedicated to an important woman in history.

While feminism was making waves among students, activists, intellectuals, artists, and designers, housewives remained, for

Conversation pits saw sofas ready for intimate encounters, garish patterns and wallpaper abounded, as did a nature-inspired color palette spanning reds, oranges, browns, yellows, and greens.

the most part, tied to the kitchen. For those leaving the home to undertake paid work, appliances became a necessity as much as they were a novelty. The 1970s saw Krups, SEB, Kenwood, Rowenta, Melitta, Siemens, and National enter the home-appliance market, and domestic life was enhanced by the advent and popularization of a number of products. In the bathroom, blow-dryers and dental brushes proved instant hits, while in the kitchen, high-tech appliances inspired new kinds of cooking.

The food processor had its debut as a compact domestic appliance in Paris, 1971 as Le Magi-Mix by inventor Pierre Verdon. Brands saw the revolutionary potential of the design, and quickly followed with their own versions of the concept, making the food processor ubiquitous in kitchens around the world. Coffee had become increasingly popular and so too had home coffee machines that integrated new technologies and materials with fashion-forward styles. Whether it was the Krups T8 (1974), the Krups Duomat (1976), the SEB Cafetière Filtre (1979), or the Philips HD 5113 (1973), these new appliances meant that hosts could easily make a party's worth of servings to caffeinate their guests.

Freezers, which became widely commercially available in the 1940s, only became commonplace in homes during the 1970s. Once established, their presence in kitchens was revolutionary, helping those who no longer had time to shop and prepare food every day. Dishes could be batch cooked, frozen, and eaten at a later date, and shoppers could stock up on pre-prepared frozen vegetables and meat—once more liberating women from a task historically tied to the role of housewife. This changed what families were eating, too. Seasonality became a thing of the past, as freezing meant that all vegetables could be eaten all year

The 1970s breakfast bar in the kitchen made eating a casual occasion.

Domestic life was enhanced by the advent and popularization of a number of products.

round. More excitingly, ice cream was transformed from a rare treat to a household staple.

The 1970s was also an era in which food science exploded, paving the way for more convenience food and a spectrum of new and exciting tastes. Chemists began engineering flavors, such as smoky bacon and prawn cocktail, which made potato chips the moreish mouthful they are today.

An onslaught of self-care appliances from these companies created the look of the decade—you could argue that the '70s was the decade of the blow-dryer. "Haircare products had their biggest increase in the '70s, but the markets were already saturated at the end of the decade," says Jaro Gielens, interaction designer and collector. "From then on, all blow-dryers were basically the same device (and sets of attachments) in a different color setting.

As a result, the fashion industry, and hairstyles of the men and women on TV and in the cinemas, would never be the same again."

The microprocessor—the foundation of today's computers—was introduced in 1971. The era also witnessed the invention of the cell phone and the founding of Apple. Technology was starting to play an important role in people's lives. Color television became a must-have, as did videos and VHS players. Runners could break a sweat to music on their portable cassette players, and office workers could save their progress on a floppy disk—which got its name from its vinyl casing being flexible and "floppy."

Despite the period often being described as a time that taste forgot, a renewed love for the decade—warts and all—is taking place in fashion and interiors today and with good reason. In terms of domestic transformations, the '70s was one for the books.

Multipress

Braun | Model No. MP 50
Germany, 1970

This fruit juicer is a follow-up to the MP 32, which was released in 1965. The design was upgraded in 1970 by Jürgen Greubel, who worked under Dieter Rams between 1967 and 1973, primarily on household and body care appliances. The MP 50 is the best known version of the Braun juicer range; many German children of the 1970s will remember it from their childhood kitchens. Its pieces lock together when its metal bar is lifted into position. Pulp is captured in the top container, and juice flows out of the nozzle below the logo. After leaving Braun, Greubel went on to work at the Design Research Unit in London, collaborating with Transport for London, among others.

Biomaster

Krups | Model No. 251
Germany, 1976

If this product looks like the Multipress, it's because they are near carbon copies of one another. They are the same size, have the same cylindrical form, and each has a small spout in the middle. Both products include a metal element to hold the body and lid in place, while acting as a handle—making them easily portable. Both can juice fruits and vegetables, and sit supported by sturdy plastic feet designed to reduce vibration. It's unlikely that the average consumer would have noticed a difference between the two. The Krups Biomaster, a rare product to get a hold of today, was even on the market for the same price.

Lady Braun Super Hairstyling-Set

Braun | Model No. HLD 50
Germany, 1972

This product promised and delivered an all-in-one hairstyling solution for women. The blow-dryer came with several attachments, including a handy spray-bottle fixture that could dampen hair in one easy spritz while blow-drying. A curling wand was also part of the package. This type of styling set would be imitated by several competitor brands, but Braun pushed on and released the HLD 80 in the same year, a supersized version with even more attachments. A very similar styling set for men was also released in 1972 (page 46).

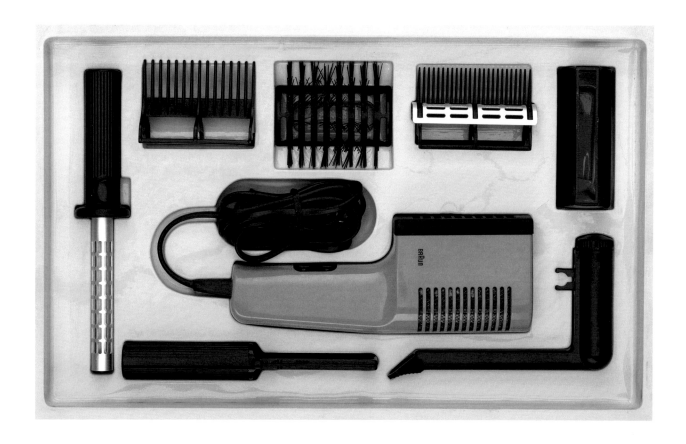

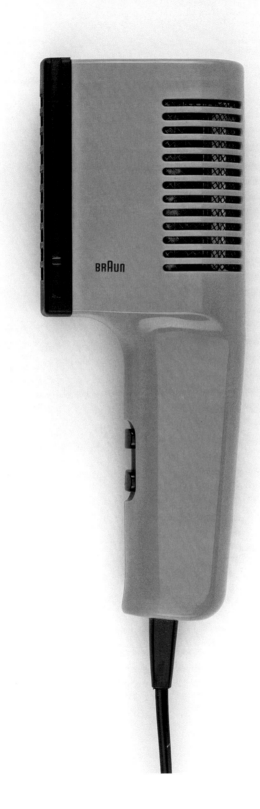

Braun worked with natural-looking models that would serve the products credibly.

Lady Braun®
Super Hairsty

Neu: Wasserzerstäuber

Zum effektvollen Auffrischen der Frisur.
Die gleichmäßig angefeuchteten Haare können
leicht und haltbar geformt werden.

g-Set

BRAUN

Das komplette Styling-Set

Zum Trockenkämmen, Trockenbürsten, Formen, Wellen, Entwirren, Glätten, Auflockern und Fülle geben.

u: Ondulierstab

n Locken
zelner
arpartien.

Lady Braun Super Hairstyling-Set | 63

Foen 1000

AEG
Germany, 1976

This handsome blow-dryer was part of the same line as the Foen 500 but could easily be mistaken as a Krups or Braun product. In fact, the sturdy, glossy plastic shell is not typical of AEG, which veered more towards matte plastics. The handle can be tilted along the axis to create better angles for easier hairstyling. AEG originally obtained a copyright for "foen," a German word for blow-dryer, meaning other brands had to get creative when naming competing hair drying products.

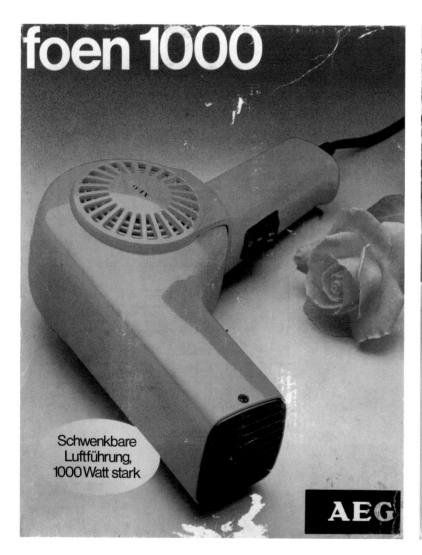

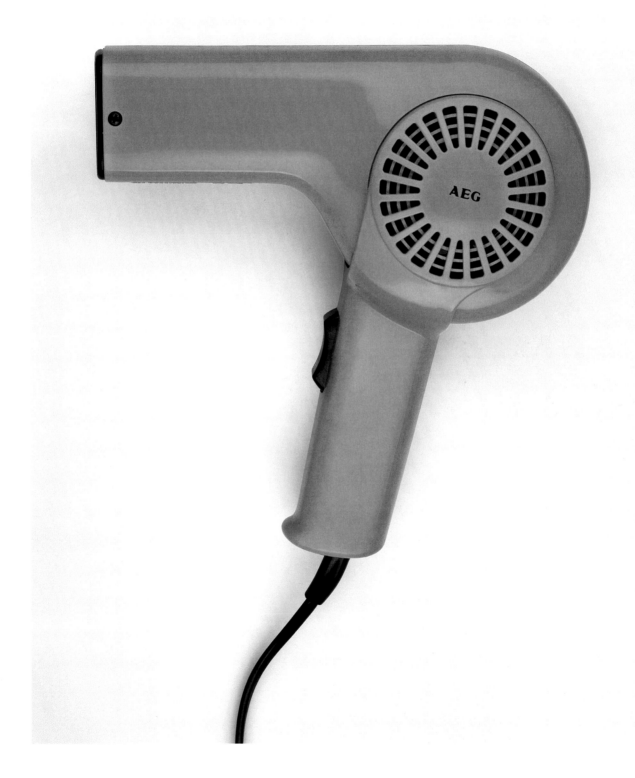

KM 40

Krups | Model No. 208
Germany, 1976

Krups was in the market of manufacturing and selling coffee grinders from the early 1960s. By the mid-1970s, coffee grinders had made considerable progress in terms of both safety and strength. The KM 40 isn't the most advanced product, but therein lies its charm. This was a straightforward and affordable product, whose mechanism functioned through a simple, rotating blade. To stand out among its competitors, Krups opted for a diagonal line where the lid fastens to the body of the grinder—a subtle, yet distinct detail.

Aromatic

Braun | Model No. KSM 1
Germany, 1967

The Aromatic coffee grinder would sit just as comfortably on store shelves today as it did back in the late 1960s. Designed by Reinhold Weiss, the coffee grinder is minimal and pared back, a hallmark of Braun's output at the time. Available in several fashionable colors, including mint green, black, and white, the Aromatic's shell is made of slender plastic, while its activation button is softer, fashioned in a contrasting color. The grinder mechanism works with a rotating metal blade, and the plastic lid can also be used for measuring and decanting beans. The functional design of the coffee grinder transcends time. Today, the device is a collector's item.

Ladyshave

Philips | Model No. HP 2111
Austria, 1976

In 1976, this compact product hit the market in an all-new color scheme. The teardrop shape proved to be a stylish design for the 1970s, softened around the edges with round corners. No flat surfaces here. Its turn-to-open carrying case was much more economical in size than the previous Special Ladyshave, with a lid and base that fit together like puzzle pieces. Designed by Murray Camens, this product was manufactured in Austria, where Philips had specialized factories dedicated to developing micromotors, which were also used in can openers and knife sharpeners.

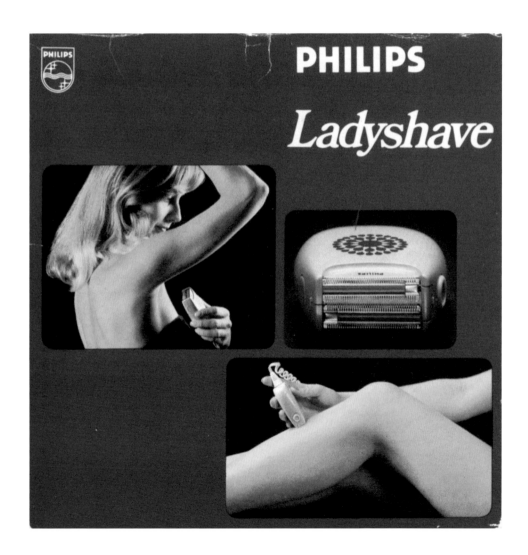

Lady Braun Cosmetic-Shaver

Braun | Model No. 5650
Germany, 1972

This women's razor is one of Braun's iconic twentieth-century designs. The product, which was designed by Florian Seiffert, who worked in Braun's design department between 1968 and 1973, is clearly a descendent of the Rams's S60, another Braun razor that's still revered today. The Lady Braun came with its own soft plastic storage case, with two compartments for the spiral cable and the device itself. At the time, the only real competitor in the women's electric-razor market was made by Philips.

Lady Braun®
cosmetic-shaver

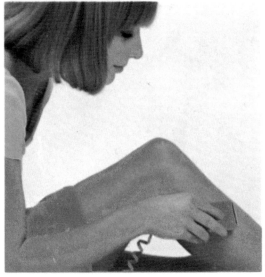

BRAUN
Made in West Germany

Less, but better. The German company that proved a minimalist approach could yield maximum results in the realm of home appliances.

Braun

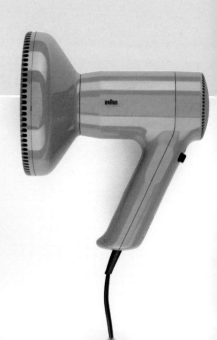

Dieter Rams, the design director of Braun from 1961 to 1995, photographed in his office in Frankfurt, Germany in 1979.

A group of men and women sporting the Lady Braun Luftkissen-Trockenhaube, a bonnet hair drying device, released in 1971.

Synonymous today with timeless design on a global scale, the esteemed personal care and household electronics brand Braun has humble roots in a small workshop in Frankfurt am Main. The company originally began in 1921 as a producer of radio parts, established by mechanical engineer Max Braun. By 1929, the company had become so successful that it began producing complete radios, becoming one of Germany's leading manufacturers.

During the Second World War, most of Braun's factories were destroyed. After the conflict, Max Braun began rebuilding and expanding the business with renewed energy. In 1950, he released the Multimix kitchen blender, and the trailblazing S50 foil shaver (facial hair was apparently his pet peeve), devices that would eventually establish kitchen and bathroom appliances as two core segments of the Braun brand.

When Max Braun died unexpectedly in 1951, he was succeeded by his sons Artur and Erwin, an engineer and business graduate respectively. They hired the German theater set designer Dr. Fritz Eichler as a cultural adviser and set about modernizing the brand. Braun soon began a fruitful partnership with the Ulm School of Design—the new successor to the Bauhaus school—alongside various other external designers and consultants, with a view to conceiving a new line of mass-producible household appliances for a better, brighter future. They were inspired in this mission by the Bauhaus's simple, human-centric design.

In 1955, a young architect and former Ulm tutor named Dieter Rams was enlisted to redesign the company's offices and was employed as a product designer the following year. In 1961, Rams became head of Braun's newly established design department, one that would soon become synonymous with pioneering German industrial design. Rams coined the famous Braun tagline Weniger, aber besser ("Less, but better"), an embodiment of his belief that design should eschew trends, centering instead on the essential aspects of form and intuitive functionality.

Experimentation was widely encouraged, and throughout the 1960s, Braun's team awakened consumers to the revolutionary potential of well-designed technology in everyday life. Products at this time ranged from breathtakingly innovative radio designs, such as Rams's sleek T 1000 Weltempfaenger (1963), the first radio to receive all frequencies available worldwide, to Reinhold Weiss's strikingly minimalist KSM 1 (1963) coffee grinder. New possibilities were now just a click away.

In the 1970s, the widespread influence of pop art took hold across the design world, evidenced by Braun's embracing of bright colors and playful shapes—within the parameters of elegant functionalism, of course. Look no further than Jürgen Greubel's Lady Braun Luftkissen-Trockenhaube (1971) for example: the now-iconic, satsuma-hued bonnet hair dryer. Or Florian Seiffert's Lady Shaver (1972): a cylindrical battery-operated razor, which saw three functions activated at the slide of a single button.

The 1980s saw Braun set a new standard for clocks and watches, defined by clarity and refinement, while the brand continued to hone its range of forward-looking kitchen and bathroom appliances, spanning everything from compact food processors to ergonomic hair dryers. In 1984, Braun became an outright subsidiary of Gillette (the American company had owned a controlling share of the brand since 1967), which in turn was purchased by Procter & Gamble in 2005 and thus continued to skyrocket. Rams left Braun in 1995, but not before defining his legendary 10 principles of good design, a blueprint for his successors.

Throughout these endeavors, Braun has retained its reputation as a creator of cutting-edge products, even as it has grown, remaining a world leader in multiple categories of small domestic appliances—each characterized by visual simplicity, enduring longevity, and human-focused functionality.

Braun 550

Braun | Model No. HLD 550
Germany, 1976

There are no bells and whistles in the design of this mid-1970s blow-dryer. A slim, pure design, its only flourish is its bright orange color—a testament to its decade, more so than anything else. The device has a unified shape, with a small area of neatly positioned ventilation holes. Its power cable can be wrapped up and stored in the dryer's handle. The product design is by Heinz Ulrich Haase, who worked with Braun between 1973 and 1978 and designed several blow-dryers during that time.

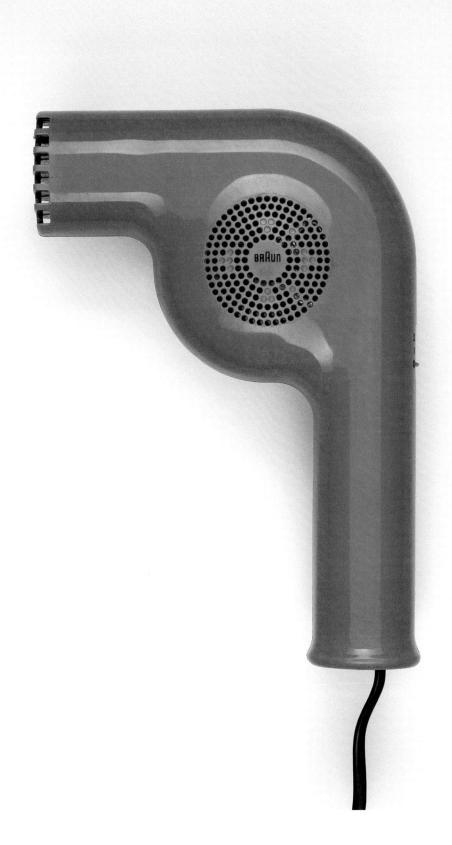

Citromatic de luxe

Braun | Model No. MPZ 21
Spain, 1972

The Citromatic de luxe was designed in collaboration between design lead Dieter Rams and full-time Braun designer Jürgen Greubel. It was intended specifically for the Spanish market, just after Braun acquired the Spanish company Primer and began manufacturing at their Barcelona facilities. The product was so successful that it was sold all over Europe. In 1994, Braun released a slightly updated version of this product, the MPZ 22, also designed in collaboration between Rams and Greubel.

Coffina Super

Krups | Model No. 223
Germany, 1976

Krups released a series of coffee grinders in the Coffina family, but the Coffina Super, designed by Krups's long-standing lead designer, Hans-Jürgen Precht, is the largest and took on an unlikely trajectory. Its retro-futuristic design inspired Hollywood prop designers for science fiction films like *Back to the Future* and *Alien,* in which it makes an appearance. Hollywood aside, this product marked a significant moment for the German brand Krups, distinguishing it as an equally-capable competitor of Braun.

Coffee Grinder

Philips | Model No. HR 2109
Netherlands, 1970

The HR 2109 was an inexpensive, sturdy, simple, and classic coffee grinder available in red, yellow, and white. Designed by Peter Stut, the product marked a departure for Philips, whose coffee grinders from the previous 15 years often seemed dated.

The HR 2109 shifted their design output in the coffee grinder category, making it more in line with what their competitors were producing. Looks aside, this product is good quality and straightforward to use.

Kaffee-Mahlwerk

SHG | Model No. MK 521
Germany, 1978

This large rectangular design is unusual for a coffee grinder, even one from the late 1970s. It did, however, offer ample space for coffee bean storage in the top compartment, with two additional transparent windows in the machine: one to see the beans move towards the grinding mechanism, the other in the clear compartment collecting the coffee grinds. In this respect, there's an almost architectural quality to this machine. It's a good example of a high-quality piece of design produced by a small company.

Supermax Swivel

Gillette | Model No. 3950
Japan, 1977

Although released by the American brand Gillette, who acquired Braun in 1967, this blow-dryer was manufactured in Japan, and it's not immediately clear who designed it. This is a shame, as it is a unique design that could easily be a decade younger than it really is. The Supermax Swivel is simple, geometric, and features a bold pop of red. The dryer can turn upwards into a vertical position, where a brush attachment can be used to style voluminous 'dos. The dryer also comes with a chunky plastic comb in the same vibrant red.

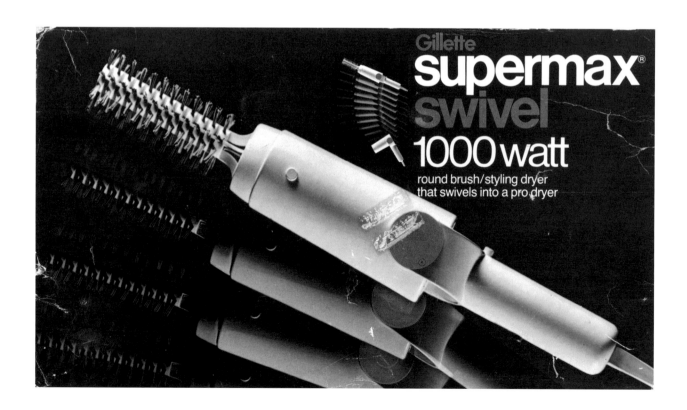

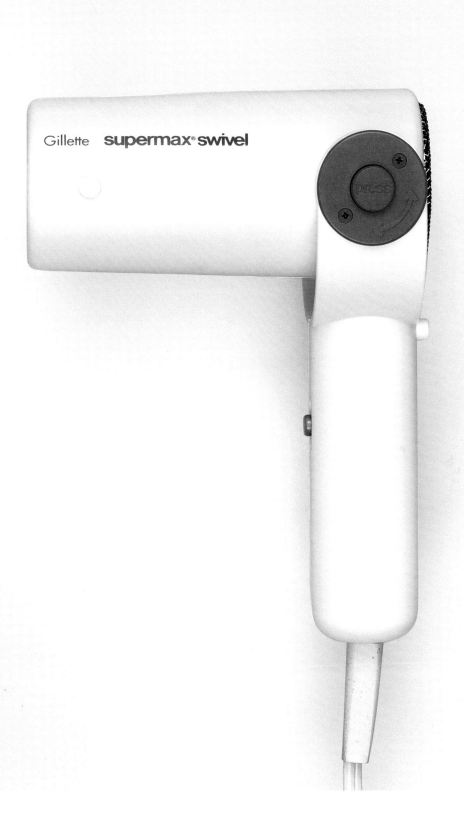

Addigramm M

Krups | Model No. M 844
Germany, 1975

Before Krups produced coffee grinders, espresso machines, and juicers, it made kitchen scales. Founded in 1846, the brand began life making precision scales. When the Addigramm M was released in 1975, Krups was Germany's leading scale manufacturer, known for its quality and accuracy. The gleaming white plastic piece comes with a matching mixing bowl, which doubles up as the weighing plate when placed on the scale. Without the bowl, the scales are free to weigh meat and other ingredients. When it's not in use, the bowl can be flipped upside down and placed atop the scale like a lid—a considerate, space-saving feature.

Intercity

Braun | Model No. 5545
Germany, 1977

This product is clearly targeted to business travelers. Even the name, evocative of the Intercity train services launched by the German rail network Deutsche Bahn in 1971, leaves little room for confusion. This handsome device comes with a special dock that can be wall mounted for charging, while the razor itself can easily be packed away for the busy man on the move. The all-black device matched well with other Braun designs released that year, including the Braun ET66 calculator (also in black) and the DW 20 wristwatch.

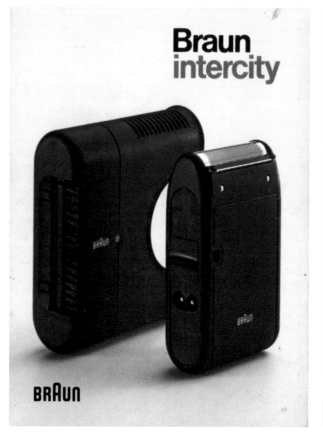

Thermic Jet

Krups | Model No. 439
Germany, 1976

With a futuristic name alluding to aviation and spaceflight, this humble blow-dryer would go on to become a mainstay of German bathrooms throughout the 1970s. This model was the latest and best in a long range of blow-dryers released by Krups, which, with every iteration, became closer and closer to the design of Braun's blow-dryers. Its soft rounded shape conceals a relatively large motor compartment, and the dryer came with a bracket to affix the device to a wall or hold it upright on a counter. The Thermic Jet was available in three colors, orange (pictured), white, and avocado green. It remains the largest, most powerful and, arguably, among Krups's best-looking blow-dryers of all time.

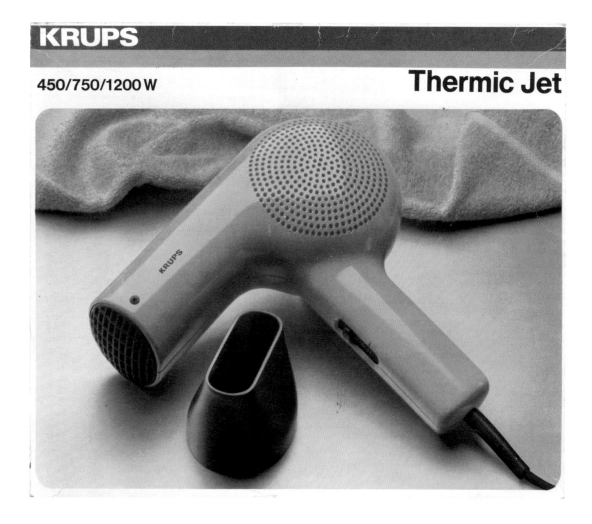

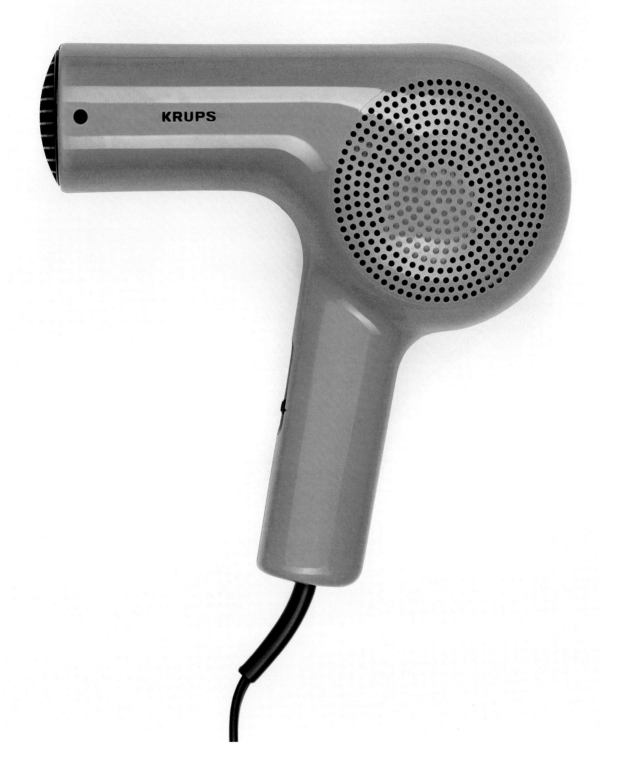

Die Schwebe Leise

Philips | Model No. HP 4628
Netherlands, 1978

This chartreuse device wouldn't look out of place in a dentist's office, but was in fact a product that promised both freedom and well-dried hair. The by Peter Nagelkerke designed HP 4628 was the *crème de la crème* of bonnet dryers. The name translates to "the silent hover," a quality afforded by the motor being tucked away and carried in a shoulder bag. While this meant less noise around the ears, it also meant greater multitasking for the wearer, as evidenced by the model on the packaging cheerfully engaged in other pursuits, such as talking on the phone, tidying her home, and taking care of her children. The device's solid, hard-shell plastic has a similar quality to another revolutionary Philips product line of the era, the all-plastic HR 6240 series of vacuum cleaners.

Advertised as low-noise, quick, and convenient, "Die Schwebe Leise" was the best bonnet dryer on the market at the time. Packaging photography by Christopher Joyce and packaging design by Henk Jan Drenthen.

Power Turbo

General Electric | Model No. PRO-10
Singapore, 1977

For something as quotidian as a blow-dryer, the Power Turbo has a menacing quality to it, no thanks to its name, which is emblazoned on the barrel in the chunky retro typeface that General Electric favored at the time. The design takes the idea of a "pistol dryer" a step further with its shape. Resembling a revolver, its grip has divots and its on-off switch looks like a trigger. The device has three drying settings and is coated in a sturdy beige plastic.

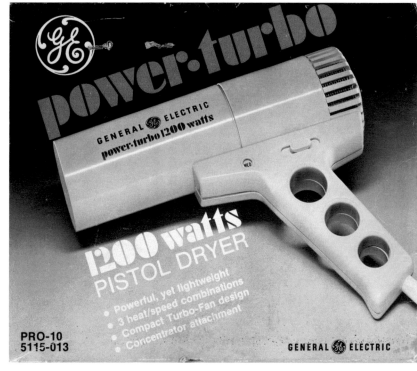

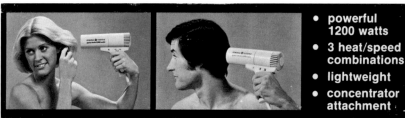

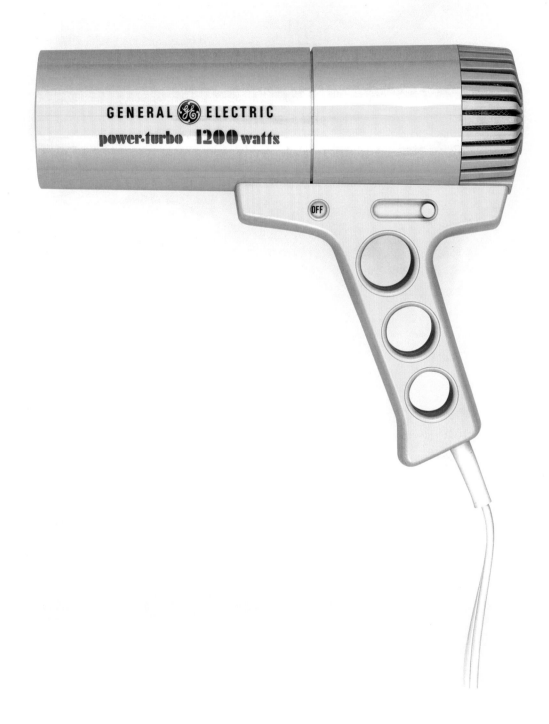

Automatic Egg Boiler

Tefal | Model No. 39907
France, 1976

Tefal's mid-1970s egg boiler followed a typical design: a charming little circle where eggs fit together snugly as though back in the nest. It's made of a robust plastic shell in '70s orange, with a transparent brown covering and the logo emblazoned on the lid. The one slight difference with this version is that it has a handle in the center, rather than the more common placement on the side. Other brands sold this same product in a selection of colors.

Automatic Egg Boiler | 99

The Looking Glass

General Electric | Model No. IM-4
Japan, 1978

"Versatile!...Convenient!" exclaims the box of General Electric's Looking Glass mirror. These promises are the second element you'll notice on the packaging, right after the retro, chunky lettering. The enthusiastic messaging pertains to the three ways that this simple geometric mirror can be configured: as a handheld device, on the countertop, or wall mounted. With a removable handle, the mirror can easily be packed away for use while traveling. Contained in a hexagonal frame with bright yellow accents, the circular mirror casts a ring of light on both the standard and the magnified mirror on the reverse. The Looking Glass was a copy of the infamous Consul Spectralux, a mirror designed by Dieter Rams and produced in Braun factories in Barcelona, Spain.

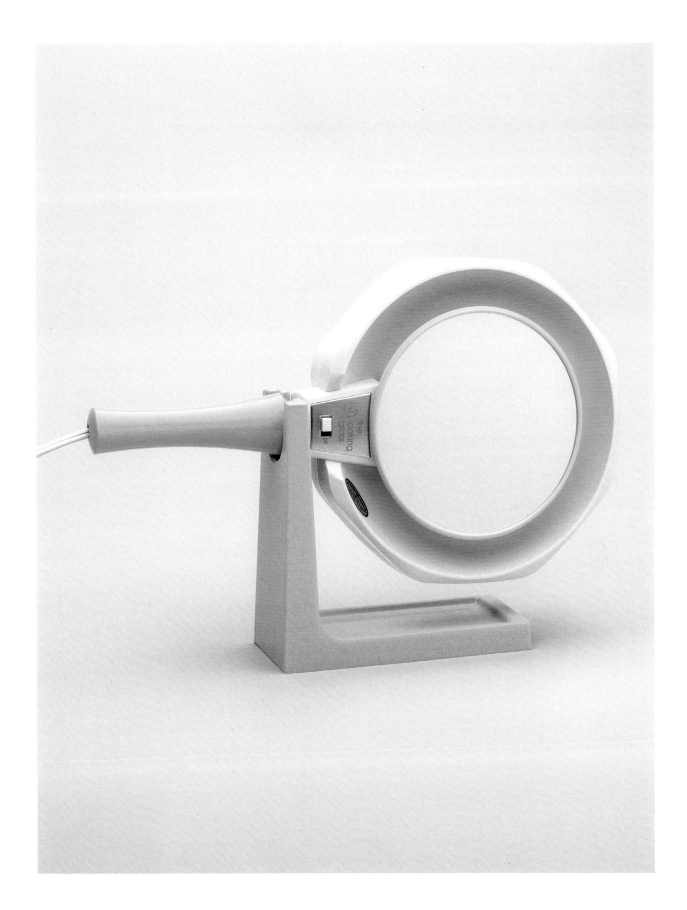

Elektro-Filterkaffeemühle

Emide | Model No. KM 11
Germany, 1978

The design of this 1970s coffee grinder doesn't immediately indicate its multiple applications. Removing the lid exposes the grinding mechanism, revealing the device's additional function. Able to accommodate Melitta 101/102 paper filters, the lid, when turned upside down, can be placed atop a mug or a coffee pot to make pour-over coffee. Another nice touch is that the Emide logo is written twice, horizontally flipped, making it legible when the device is used for grinding or making coffee. The KM 11 also provides another surprise, the uncommon feature of having a pneumatic timer instead of a simple mechanical one.

Elektro-Filterkaffeemühle

Mahlwerk-Kaffeemühle K4

Bosch | Model No. K4
Germany, 1974

This compact grinder was introduced in 1974. Its edges are soft, and it has a clear plastic compartment for freshly ground coffee. The K4's controls are also emphasized with color. The oversized on-off button and a simple switch that toggles between grind settings stand out from the device's plastic body. This product, which was available in four colors, is a marked evolution from the K1, released in 1968, where the motor block is twice the size, as well as the 1974's K3, a popular, nearly identical product, which was sold by several other brands also.

Mahlwerk-Kaffeemühle K4

Kaffeemühle K12

Bosch | Model No. K12
Germany, 1977

Upon release, this was among the best coffee grinders on the market. Its simple design lets users grind fresh coffee directly into a filter, with several settings determining the quantity and coarseness of the grind. Thanks to a generously sized container for storing coffee beans, the machine could easily—and accurately—grind beans for up to 12 cups at a time. The K12 was available in three colors, and a similar product with a more rectangular form was sold by rival Melitta.

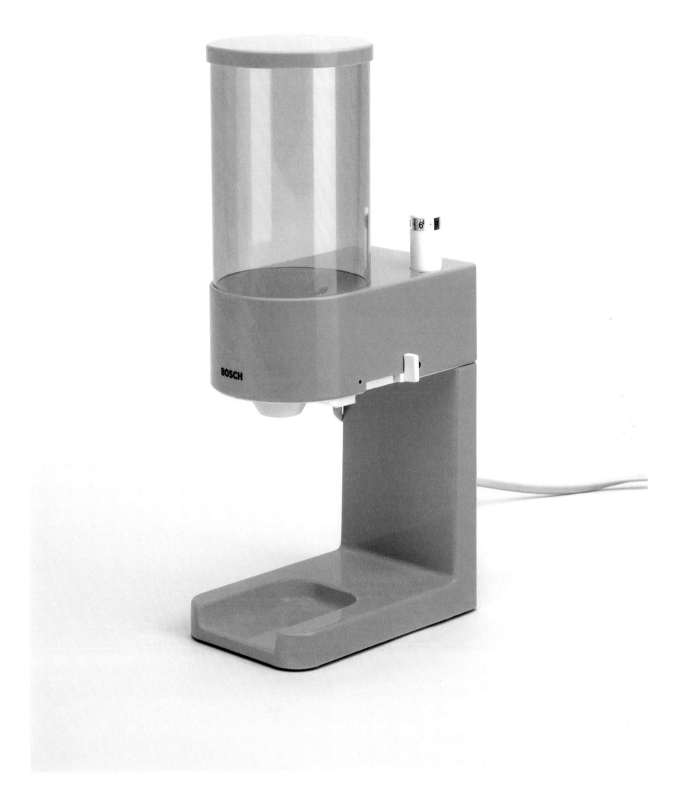

Kaffee-Automat T8

Krups | Model No. 265
Germany, 1974

This is an iconic coffee maker by Krups that offered consumers a different way to brew their coffee at home. Unlike standard filter machines, where water slowly trickles into the filter, the T8 uses a pressurized mechanism, where boiled water is pushed up to the top of the machine and brews coffee in one short action. The result is a decent cup of coffee and a much shorter brewing time. An updated version of this machine, which was originally designed by Dieter Weissenhorn, was re-released in the early 1990s with a simpler construction but less durable build.

Cafetière Filtre

SEB | Model No. 980
France, 1979

SEB's Cafetière Filtre was futuristic, with round corners and pleasant curves. By this point, SEB had considerably improved their manufacturing techniques and were able to work with sturdy plastics that had glossy finishes. This is also one of the earliest examples of a feature that comes as standard today: the drip stop. SEB was among the first manufacturers to introduce it, allowing the coffee pot to be removed mid-brew, the machine designed to know when to stop the flow of liquid.

Kaffeemühle

AEG | Model No. KM 101
Germany, 1979

This AEG coffee grinder is an object of handsome proportions, the one shown here being the rare dark-brown model. Its simple form, which features a smaller box stacked on top of a larger one, is both intuitive and easy to navigate. Rounded corners soften the linear shape. This clever device features automatic portion control, a helpful feature for amateur coffee drinkers unfamiliar with the correct amount of coffee to grind, helping avoid unnecessary waste. The storage container holds up to 250 grams (8.8 ounces) of coffee beans. Electric drip coffee makers using ground coffee from a device like this became mainstream in the 1970s, popularized by brands like Mr. Coffee in the United States.

Kaffeemühle | 113

Solitair

Krups | Model No. 466
Germany, 1974

The Solitair was expensive, but it offered the gift of two products in one. Designed by Rudolf Maas, the compact, yet surprisingly powerful hair bonnet could be detached from the motor, which doubled as a standard blow-dryer. It was sold with a soft plastic carrying case for easy portability.

Because the motor could be worn on the chest, suspended like a necklace, it gives the wearer freedom to move around the house. This feature prompted the West German authorities to compel Krups to include a safety leaflet warning users not to wear the Solitair while bathing.

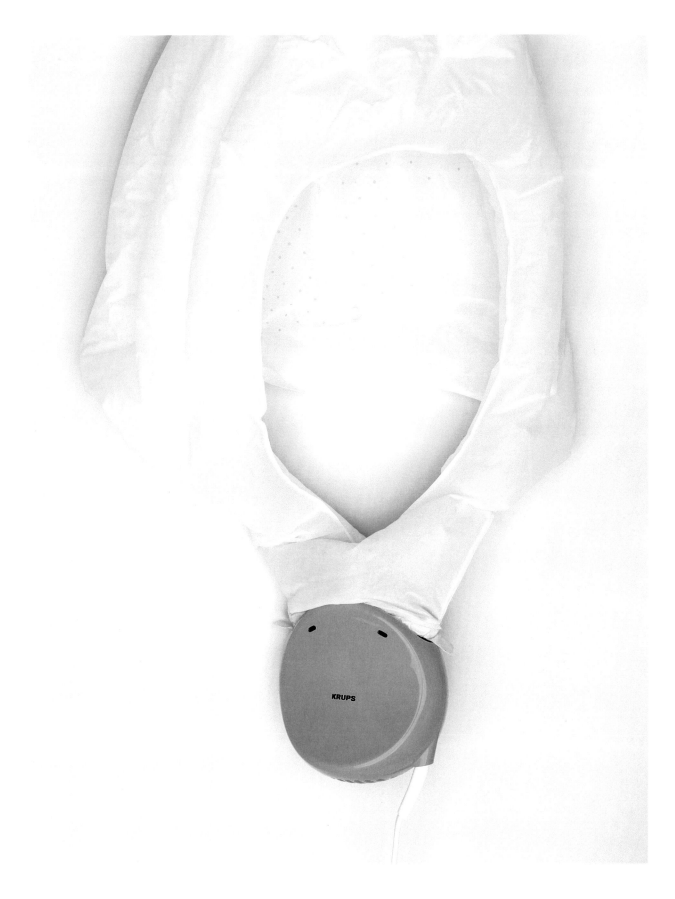

Joghurtgerät

AEG | Model No. JG 101
Germany, 1977

The yogurt maker, like the egg boiler and the fondue set, was a common must-have for most middle-class kitchens in the 1970s. It was made with a starter culture and milk, and was mixed, as is the case with this model by AEG, in six portions in individual glass jars. This enduring design, which slowly makes yogurt over a course of several hours, would remain largely unchanged in future iterations. Although the yogurt maker has dwindled in its popularity, a product like this would no doubt still come in handy today. The fresher the yogurt is, the healthier it is; plus, making it at home drastically reduces plastic waste.

Joghurtgerät

Super Mijoteuse

Tefal | Model No. 150
France, 1978

Tefal's Super Mijoteuse combines a round cooking pot with a square design. This wasn't the first slow cooker on the market, but it was among the first with a digital timer and an incorporate single-digit LED output. Its analog elements are graceful as well, particularly its orange switch that toggles between temperature settings. This modern device offers cooking times of up to nine hours. An earlier version of this product had a clear lid, which was replaced by tinted glass, which was slower to show signs of aging.

Futura Electronic 4109

Moulinex | Model No. 377
France, 1977

The Futura Electronic remains a legendary retro-design product. With a red plastic and chrome base, this version from 1977 wasn't a huge leap from the previous iteration of three years prior, largely because it didn't have to be. Rather than a design update, it's a technological makeover introducing a console-style control panel with a digital timer in a VFD display—LCD wasn't available until years later. The design is by Jean-Louis Barrault, who helped define Moulinex's design vision. He also designed bottles for L'Oréal and the off-road vehicle Méhari for Citroën, among many other contributions to French design.

Yogurt Maker

Moulinex | Model No. 508
France, 1975

Against a landscape of boxy, geometric kitchen devices, this circular yogurt maker by Moulinex is a decided departure. However, very much in keeping with kitchen products of the mid-1970s, it is clad in a bright-orange plastic shell. The yogurt maker lets the user select from different temperature settings, and includes an automatic power-off function—a helpful feature for the hours-long process of making yogurt. This allowed users to switch it on and simply forget about it until the process was complete.

Fruitpress

Moulinex | Model No. 502
France, 1976

The mid-1970s fruit juicer by Moulinex is a practical, simple design. The sunny yellow press is affixed to a white base, with a clear plastic container below to catch the juice and pour it via a small nozzle. Instead of using an on-off switch, the electric motor is activated when the fruit is pushed down onto the press. The device is tall enough to fit a glass below the nozzle, though the 502 was available with an additional attachment, a bowl, that could be fixed below the presser to accommodate larger volumes of juice.

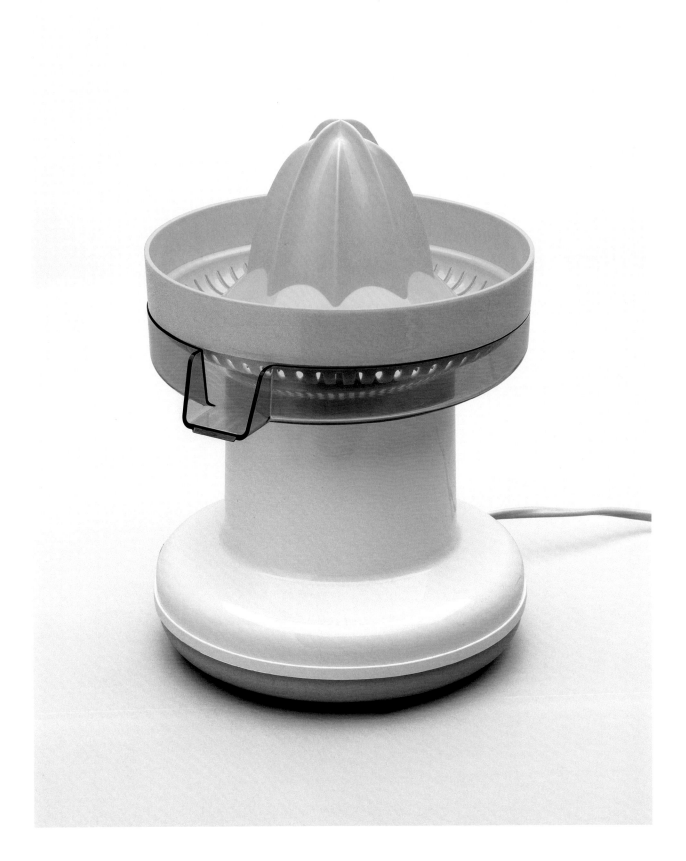

Food for thought. The French brand rendered daily chores effortless with its affordable, time-saving gadgets.

Moulinex

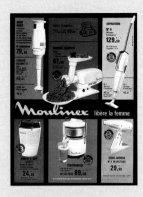

Advertising for an array of Moulinex kitchen devices, printed in November, 1963.

Moulinex founder Jean Mantelet, pictured at his desk in 1962.

The history of the famous French kitchen and home appliances brand Moulinex dates to early 1930s Paris. There, a young businessman and inventor named Jean Mantelet—inspired, legend has it, by his wife's lumpy, hand-mashed potato purée—had the idea for a food mill that would transform cooked vegetables into a smooth mash at the crank of a handle.

Mantelet introduced his device, the Moulin-Legumes ("vegetable shredder"), at the Lyon Fair in 1932, where it retailed for 36 francs but garnered little interest. Two months later, he brought the design to the Paris Fair, reducing the price to 20 francs, and achieving instant success. By the end of the year, his company, then known as Manufacture d'Emboutissage de Bagnolet, was producing around 2,000 mills a day. Mantelet had found his business model: small, mass-produced domestic appliances that were affordable, yet invaluable in their time-saving potential.

Between 1929 and 1953, Mantelet applied for 93 patents, producing everything from nutcrackers (the Mouli-Noix) to salt mills (the Mouli-Sel), and his best-seller, the Légumex: a rotating vegetable peeler and scraper. But it was 1956 that spawned Mantelet's biggest hit, a neat electric coffee grinder, dubbed the Moulinex. Retailing at less than half the market price of other coffee grinders, the Moulinex was a testament to Mantelet's theory that "a new price is a new market." By the end of the year, some 1.5 million units had sold. Capitalizing on the product's success, in 1957 Mantelet changed the company name to Moulinex, and swiftly entered the realm of electronic appliances.

The 1960s marked the arrival of Jean Louis Barrault at Moulinex, who joined as a freelance designer in 1963, fresh from Raymond Loewy's Compagnie d'Esthétique Industrielle (CEI) in Paris. Arguably the brand's most influential designer, Barrault prioritized simple efficiency—"a juicer must squeeze citrus fruits, that's all," he told Les Echos in 1996—and innovation ("innovation must be permanent otherwise you are dead").

This approach is evidenced in his multifarious designs for the company over the subsequent 25 years.

The 1960s also saw Moulinex launch its famous slogan "Moulinex libère la femme!" ("Moulinex liberates women"), targeting a new generation of housewives keen to free themselves from the daily drudgery of household chores. New releases spanned food processors of all kinds, including choppers, mixers, and blenders, which swiftly revolutionized home cooking, through vacuum cleaners and hair dryers.

By the 1970s, it was estimated that every household in France had, on average, four Moulinex products, while around 50 percent of the company's sales were now overseas. And the company's creativity, it seems, knew no bounds. Over the course of the decade, Moulinex introduced a stream of new products, from its 1972 electric coffee maker—a resounding success—to its first microwave oven in 1978, as well as more whimsical offerings, like brightly hued salad spinners and electric egg boilers.

The first half of the 1980s was also a time of prolific innovation, resulting in programmable coffee machines, nifty pasta makers, and a whole host of refined cooking appliances. Think: the modern, radically compact Vertical Grill (1981), featuring a motorized attachment for rotating chicken while it roasted, or the Cuitout, a progressive, multi-tasking Slow Cooker (1981), which could cook, stew, fry, steam, and boil.

In 1985, overwhelmed by competition, Moulinex found itself in serious financial difficulty for the first time. In 1991, following the death of its founder, the company purchased German home-appliance brand Krups, resulting in further financial turmoil. 10 year later, Moulinex declared bankruptcy and was subsequently acquired by longtime French rival, Groupe SEB.

Under this new umbrella, the brand has been able to continue its founding mission of making kitchen appliances available to all, while instilling the cooking process with an effortless *joie de vivre*.

Cafetière Espresso

Moulinex | Model No. 678
France, 1975

This iconic red espresso maker was one of the first to simplify complex professional espresso machines for everyday consumers. Although easy to use, it maintained the design language of classic Italian espresso machines—only in miniature. It was sold with two red plastic and glass espresso cups—worth the purchase alone—and a coffee spoon. Additional matching glassware was sold separately. This espresso maker is another example of Jean-Louis Barrault's contribution to Moulinex.

Electronic Egg Boiler

Moulinex | Model No. 567
France, 1979

For a product that doesn't necessarily need to exist (eggs will cook just as effectively in a saucepan rather than in a dedicated electronic device), this is an attractive egg boiler. This version, in shades of brown, replaced the previous orange-and-white palette. It comes with a rotating dial on its handle with different settings for how hard the user wants their eggs boiled. A stylish translucent plastic cover is integrated with the handle. The cable wraps around and tucks into the base for easy storage.

Allround Styler

Krups | Model No. 401
Germany, 1977

This extensive styling set came with four different types of brush attachments, a flat styling nozzle, and a small spray bottle. What differentiated it from competing products was an inbuilt mechanism allowing the dryer to swivel from 90° to 180°, giving it the two-in-one capabilities of both a classic pistol dryer and styling brush. This functionality is neatly concealed by the bulbous orange plastic casing, with graphic diagonal lines denoting the dryer's suction inlet. Other versions of this product were available on the market with fewer attachments, including the Allround Styler 800, which was a less powerful machine.

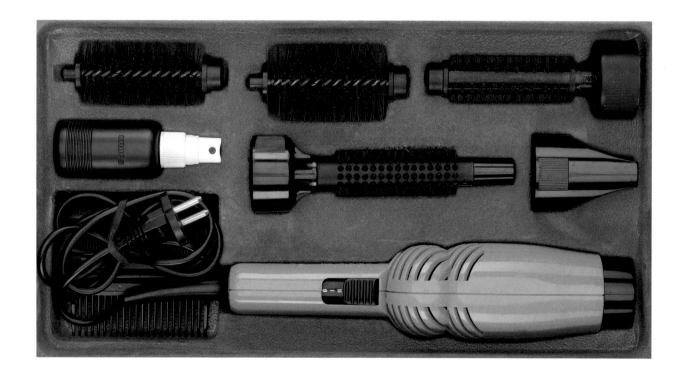

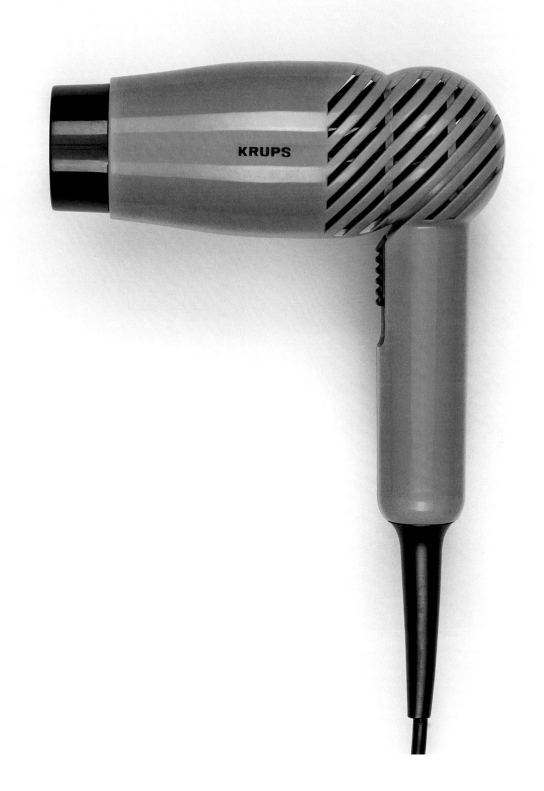

Allround Styler | 133

Popcorn Party

ITT | Model No. 12 08
France, 1979

The Popcorn Party is a relatively small popcorn machine, no larger than the size of an egg boiler. To show that popcorn isn't just for kids, the packaging depicts the snack alongside an open fire, whiskey, and a cigarette. What's more, this isn't even the most outdated part of this product. To forge a link with the perceived "Americanness" of popcorn, the French makers of this device used stereotypical imagery of Native Americans on the product. The use of such imagery by a non-native company on a non-native product remains offensive to this day.

Popcorn Party | 135

Duomat

Krups | Model No. 269
Germany, 1976

The Duomat, designed by Rudolph Maas, is an evolution of the Krups classic T6 coffee maker. In this case, the T6 is incorporated into the Duomat's symmetrical design, with the added ability to make a pot of coffee and a pot of tea—or two pots of tea, or two pots of coffee—at the same time.

Elaborating on existing designs helped create a sense of consistency across the Krups range. And in a consumer market where kitchen devices were vying to outdo one another in terms of convenience, the Duomat was a genuinely useful device for busy mornings.

Waffelautomat Luxus

Rowenta | Model No. WA-02
Germany, 1978

As far as waffle makers go, Rowenta's Waffelautomat Luxus was at the top of its class. The solid device was equipped with a special edging to ensure no waffle batter escapes, as well as a thick handle. It was perfect for making thick, round waffles with scalloped edges. The design pictured here is part of the "Dekor Grafika" collection, recognizable because of its dark-brown accents and concentric rings, which were applied to other appliances like coffee makers, toasters, and hot plates.

Elektro Messer

Bosch | Model No. EM 1
Germany, 1978

There isn't much design variation among electric knives. This model was released by Bosch in the late 1970s, and features an angled front that's similar to the classic designs Moulinex was releasing in France around the same time. It comes in dark-brown and yellow, with an on-off switch and a button to remove the blade from the handle. Users could mount this product on their kitchen walls in a special holder with separate storage compartments for the cable and blades.

Special Duo

Krups | Model No. 359
Germany, 1976

The popularity of electric knives has faded since the 1970s, but as a refresher, here's how they work. Normally, they contain two parallel blades that, when activated, move backwards and forwards, powered by a motor in the handle. While the Special Duo didn't change things in this regard, it did outperform many of its competitors. It's noteworthy that Krups included an additional, heavily serrated blade designated for frozen foods, as freezers were becoming more common in households. The use of orange accent colors and rounded edges is typical of Krups.

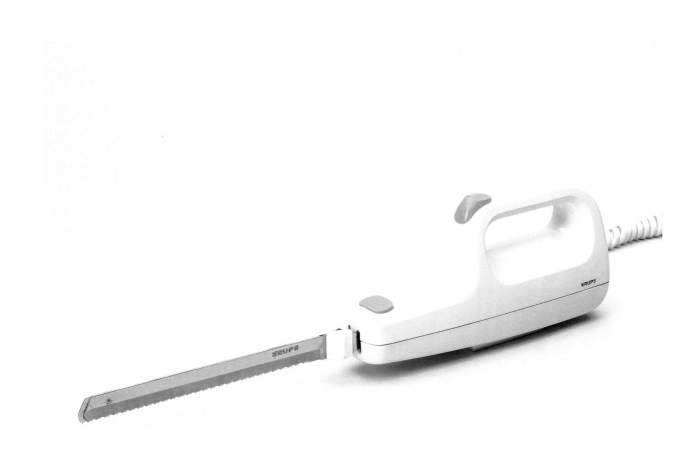

Foen Salon

AEG
Germany, 1978

The 1970s and '80s produced many blow-dryer kits with multiple attachments, and AEG's Foen Salon was no exception. This futuristic-looking modular system's defining feature is that its handle can be removed to make it even more compact, making it almost as small as a travel-size blow-dryer. With the handle removed, it can accommodate different styling attachments. Its other unusual feature is the inclusion of a cleaning brush to dust off the device and clean stray hairs.

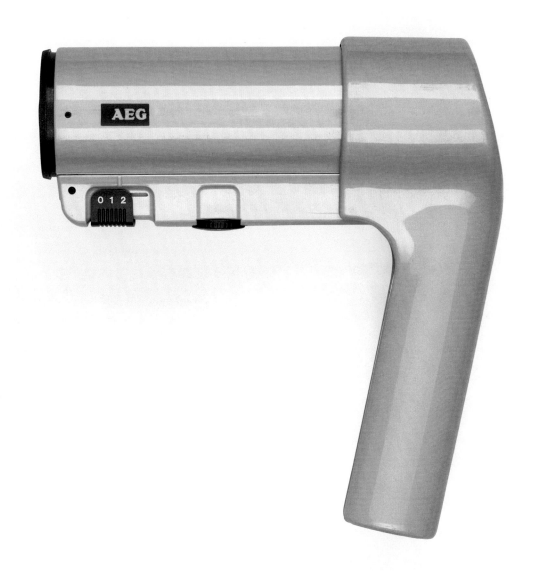

Zauberstab

ESGE | Model No. M 122
Switzerland, 1978

The immersion blender was first patented in Switzerland in 1950. It offered chefs the ability to purée or blend food in the pot or bowl in which it was being prepared. This meant no more messy transfers to a separate machine or vessel. The new appliance was named Bamix, a portmanteau of *battre et mixer*, to beat and mix. This example shows how far along this little kitchen appliance had come in 28 years. It comes with additional attachments, including blades, a grinder, and mixing jug. The original version of this design by Acton Bjorn und Sigvard Bernadotte won an IF Award in 1962.

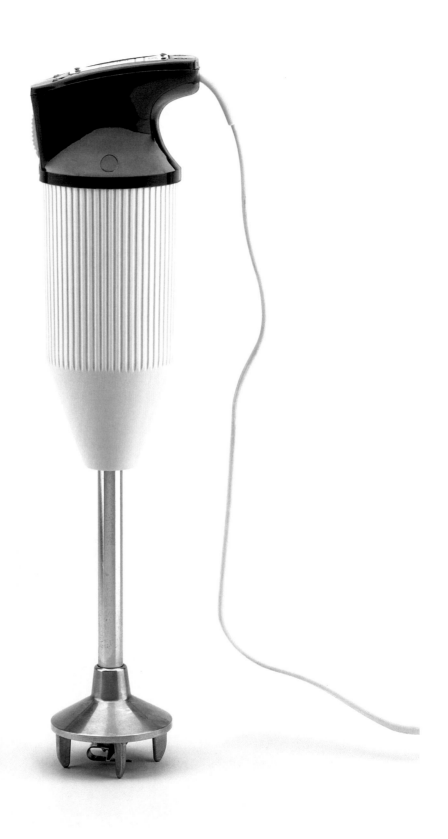

Blender

Kenwood | Model No. A 515
United Kingdom, 1978

Kenwood's late-1970s blender has an almost toylike quality, with a smooth plastic base and an ultra-simple interface. Its big, round, brightly colored power button also matches the lid. The device was available in blue and punchy yellow.

Childlike qualities aside, this was an excellent mixer using high-quality plastic. The design is by Sir Kenneth Grange (who in 1978 also designed the larger, two-speed A 520 blender), and it was manufactured in the U.K. by Thorn Domestic Appliances.

Mayonnaise-Minute

SEB | Model No. 8555
France, 1983

Contrary to what the name suggests, it only takes 30 seconds to make mayonnaise with the Mayonnaise-Minute. To operate, oil, eggs, salt, and an acid like vinegar or lemon juice, are added to the cylindrical compartment and the motor block is screwed on like a lid. Thirty seconds later, voila! The bold yellow plastic product was part of the same design language as SEB's Mini Mincer, which came with a green base and was released just a year prior. Owing to its durability and timeless design, this mayonnaise maker remains a popular, sought-after product to this day.

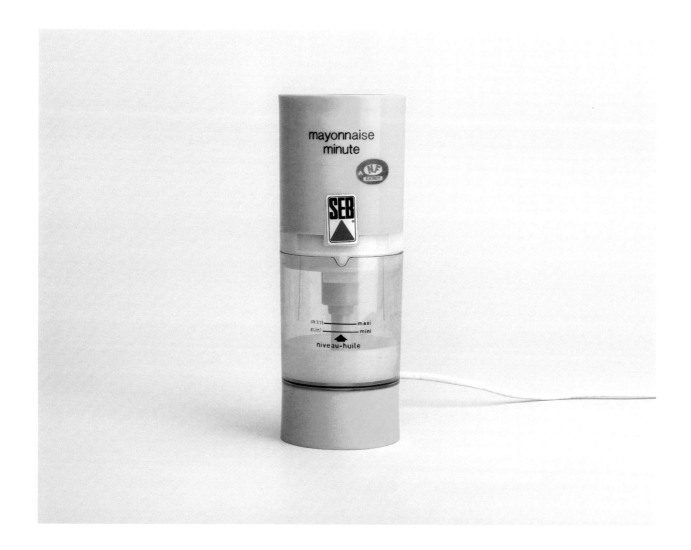

Mayonnaise Maker

National | Model No. BH-906
Japan, 1978

This is another example of a kitchen device that serves a single, highly specific purpose. National's mayonnaise-mixing machine was released in the late 1970s, coming in a sturdy plastic shell with green accents and a clear central module in which the mayonnaise was mixed by a metal beater. For a product reliant on being switched on, the BH-906's wirelessness is an unexpected choice—nevertheless, it's powered by four large C-type batteries. Composed of several parts, the device's components are easy to disassemble for cleaning and maintenance.

Cooling Bag

National | Model No. ND-101
Japan, 1978

This nifty cooling device can be worn over the shoulder like a messenger bag and has space for four small bottles to fit snugly inside. A red and a green light on the side indicate the cooling temperature. For this product to be portable, it had to be battery powered, which proved to be detrimental to its versatility, as the battery cassette takes up valuable space. The fact that it's only compatible for use with the four bottles it comes with also limits its usefulness. Despite this, what it lacks in practicality it makes up for with Western-tinged lifestyle imagery on the packaging, featuring a couple by the sea on one side and a family picnicking in front of their Volkswagen on the other.

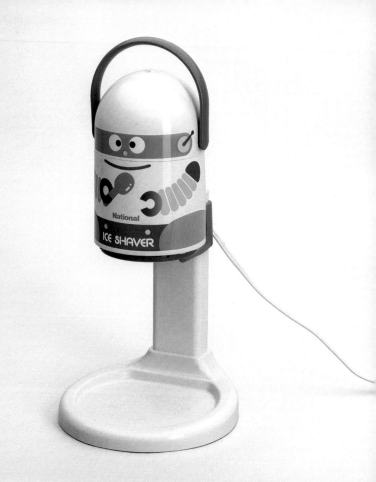

Bright ideas and a big heart. Inventive design and principles of care held the key to success for the Japanese electronics brand.

National

National Panasonic's first submersible wet/dry shaver, released in 1983.

The assembly line at the Matsushita (National) factory, Osaka, Japan in 1970.

The story of Japanese electronics giant Panasonic (formerly known as National) begins in Osaka in 1918, with Kōnosuke Matsushita, a 23-year-old inspector at the Osaka Light Company. Matsushita had devised a pioneering design for a new duplex light socket, but his boss refused to manufacture it. Undefeated, he decided to found his own company, Matsushita Electric Appliance Factory and had enlisted 20 employees by the end of the year.

Driven by the simple observation that customers would always opt for a product that was of higher quality and lower price than its competitors, he concocted a business rule to live by. Each new product must be 30 percent better than those currently on the market and 30 percent cheaper. Matsushita didn't only prioritize customer satisfaction, however. He is also credited with coining the Japanese tradition of "paternal management," whereby company staff are embraced as family from the outset and guaranteed lifelong employment. (Unsurprisingly, he is still celebrated as one of Japan's greatest business role models.)

In 1927, Matsushita began marketing an innovative battery-powered bicycle lamp under the name National—one that would become incorporated into the Panasonic brand many years later. In the 1930s, he began expanding the business to incorporate a variety of different product lines, most notably radios, which proved its biggest hit.

After narrowly avoiding closure in the 1940s, owing to the company's involvement in the war effort, by the early 1950s, Matsushita observed that the world had entered into a "new age of design," filled with endless potential. He established Japan's first-ever corporate design department, spearheaded by industrial designer Yoshikazu Mano in 1951. Soon Matsushita Corporation was producing transistor radios, tape recorders, televisions, large household appliances, and more—most of which bore the National name. This kick-started its journey to becoming one of the world's biggest suppliers of electrical goods.

In the 1960s, the company extended its output to include microwave ovens, air conditioners, videotape recorders, and countless other new products. And although Matsushita continued to produce items under the National brand, he began to sell these new items under different names too, including Panasonic, Quasar, Technics, Victor, and JVC.

The 1960s also heralded the arrival of the company's populist design approach. Aided by the burgeoning craze for molded plastics, National began creating a number of colorful and attractive designs for a youthful consumer market. Enter: the R-72S (c.1969), a battery-operated, donut-shaped radio, crafted from blue plastic and designed to be carried or worn on the wrist, and the award-winning MC-1000C vacuum cleaner (1965), a groundbreaking, easy-to-wheel design in plastic that loosely resembled a Wacky Races' car.

Throughout the 1970s, National continued to propagate playful designs inspired by the rise of pop art, and the increasing demand for products that were both functional and fun. Thanks to the brand's brightly hued appliances (like its green-lidded, easy-clean BH-906 [1978] Mayonnaise Maker and its red-pronged hair EH 161 Curling Iron [1979]), daily endeavors were rendered easy by effective, affordable design.

In 1973, Matsushita retired, but his business kept on growing; in 1974 it acquired Motorola's television operation in North America. Computers, breadmakers, and jet baths soon followed: there really was almost nothing home (or office) electronics related that the company shied away from.

In 2003, the Matsushita brand adopted the title of Panasonic as its global identity. It began phasing out its National, selling its products under Panasonic instead. Even so, the company has continued to embrace its founder's principles, creating affordable, high-quality appliances, and treating its workers with respect. In 2021, *Forbes* ranked Panasonic 194 in its poll of the World's Best Employers.

Curling Iron

National | Model No. EH 161
Japan, 1979

A straightforward product without the special features of multi-attachment 1970s styling kits, the EH 161 was nevertheless a popular curling iron from Japan. Today, much of its merit lies in its artful packaging, which offers helpful instructions with matching visuals, alongside photography of Japan's flippy '70s hairdos. This hair product was available in green or red, both with a white plastic body and a power switch in the accenting color. It also included a turning cable for easy handling.

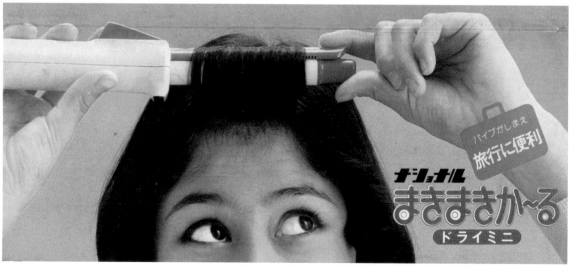

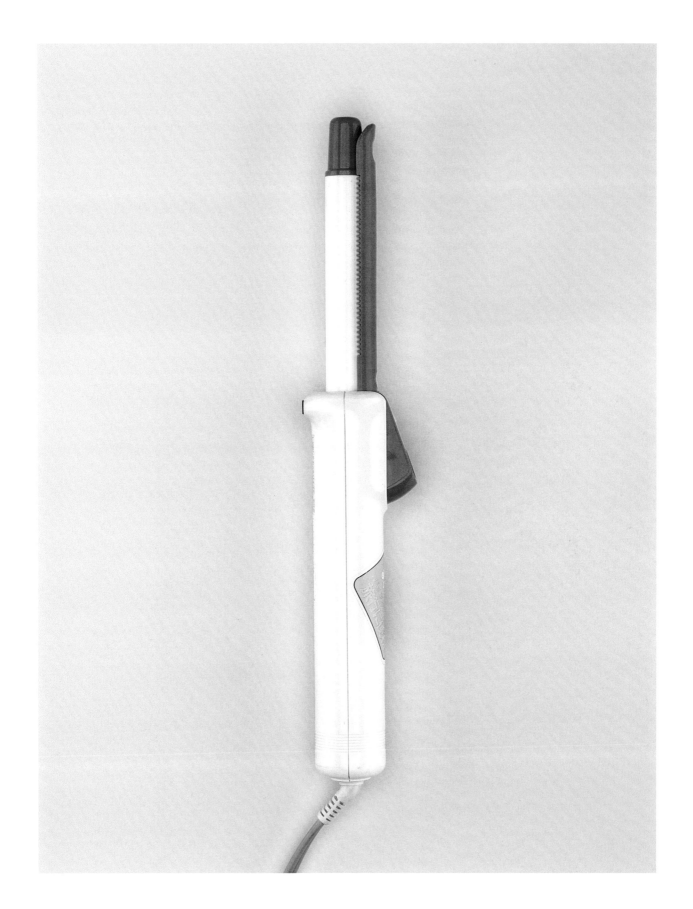

Curling Iron | 155

Speed Pot

National | Model No. NC-950
Japan, 1977

Speed Pot was a simple, no-nonsense name for a simple, no-nonsense product. This upright hot-water dispenser by the Japanese electronics brand National was easy to use and pleasing to look at. With its cylindrical tank and slender body, its design shares a few qualities with Braun's classic Aromaster coffee machine. Similar devices were sold elsewhere with equally on-the-nose names, such as the very American sounding "Hot Shot" sold in the United States by Sunbeam, and AEG's "Schnelle Tasse," which translates from German as "fast cup." This Japanese version was available in orange and green.

Sake Heater

National | Model No. NC-31
Japan, 1979

This is a beautiful product for heating sake, a Japanese rice wine that can be enjoyed hot or cold. The NC-31's compact heater combines traditional elements, such as the reeds etched into its glass, transforming them into an unmistakably modern product, powered by electricity and made of plastic and glass. Although shown here in green, the sake heater was also available in white and dark red. The packaging features a charming illustration of a man and a woman in Japanese dress enjoying their warm sake with a meal.

Cloth Dryer

National | Model No. ND-11
Japan, 1976

The Cloth Dryer could do much more than its name suggests. While its packaging advertises it as a heated drying rack with adjustable grill for dishcloths and other kitchen textiles, it also functioned as an upright blow-dryer with a nozzle that could be pointed at wet dishes and even soggy shoes. The base provides an extra extendable stand to dry larger items and can be wall mounted. All in all, the Cloth Dryer is a multipurpose drying product that is well suited to small kitchens and the often rainy Japanese climate.

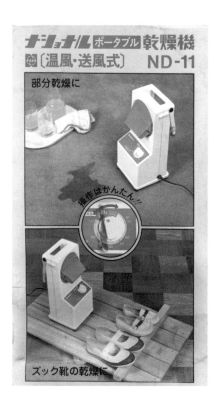
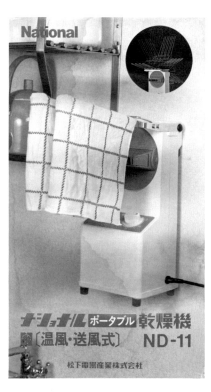
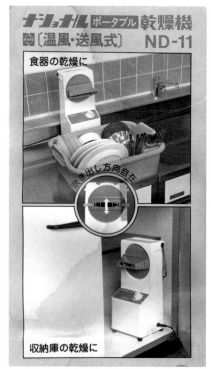

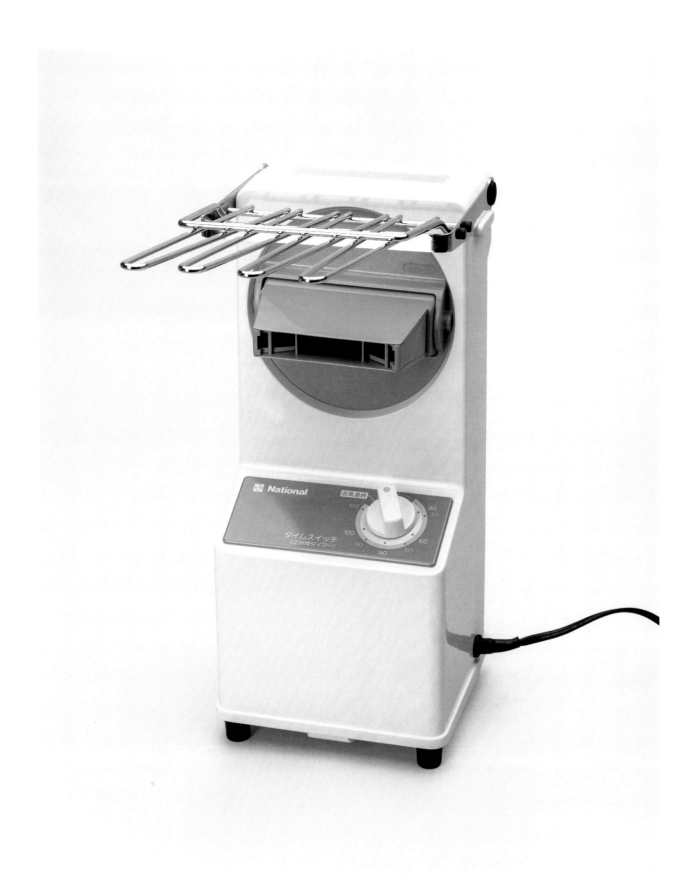

Ice Shaver

National | Model No. MF-U7
Japan, 1976

This was one of many ice shavers on the Japanese market at the time. This one would have been perfect for making *Kakigōri*, a Japanese dessert made from shaved ice, flavored syrup, and condensed milk—similar to what other parts of the world call a snow cone. Although this product came decorated to look like a cartoon robot, it was also available as a penguin, among other iterations. The MF-U7 was battery operated, so likely wasn't the most powerful product of its class on the market. The body can be adjusted for height to accommodate different-sized glasses or bowls below.

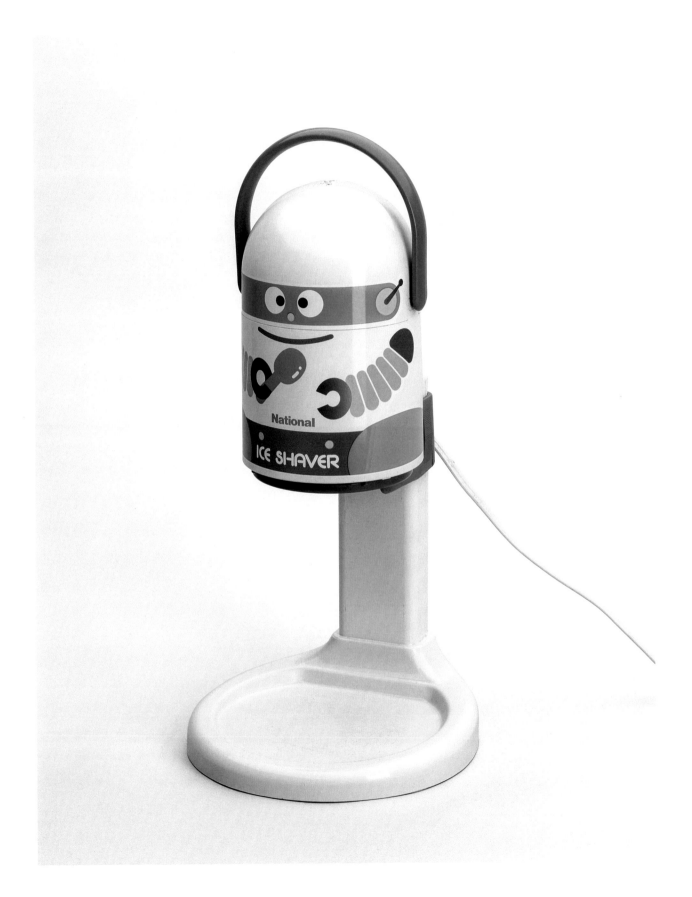

Suzette

Krups | Model No. 235
Germany, 1977

Coming in a dashing shade of 1970s orange, the Suzette was a successful gadget that made cooking crêpes easier than ever. The user simply measured the batter onto a special plate and submerged the hot surface of the crêpe maker into the batter. From there: lift, turn, and voila. The result is a perfectly thin crêpe. Following this model, designed by Rudolph Maas, there were several subsequent similar devices by other brands. None of them, however, came with the Suzette's special plate, emblazoned with a simple recipe for crêpe-making.

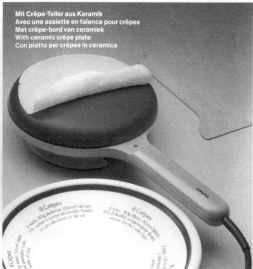

Suzette | 165

Party-Grill

Krups | Model No. 293
Germany, 1975

Kitchen products from the 1970s were all about entertaining. This grill, designed to be an interactive focal point for the dinner table, is a good example. Unlike other electric grills featured in this book, notably the Philips Rotating Grill (page 168), Krups's Party-Grill is a straightforward and safe design that functions like a plug-in frying pan. Once cool, the grill plate can be removed for easy cleaning. Available in burnt orange and canary yellow, its ventilation inlets loosely borrow from the design language of the sports car.

Rotating Grill

Philips | Model No. HD 4151
Netherlands, 1978

Designed by Hans Jülkenbeck, this product raises a few questions—and more than one is about safety. The rotating grill brings an exposed heating element that's hot enough to cook meat right to the middle of the dinner table. How was this approved? What's less helpful is that its base isn't very heavy, making it easy to tip over. Later versions featured a protective bar around the middle and, puzzlingly, the product remained on the market until the 1990s. Did anyone burn their fingers trying to retrieve their steak skewers? Or was this the type of product that sat under the Christmas tree, to be used once, perhaps twice, before going the way of the oft-forgotten fondue set?

Popcorn Center

Black & Decker | Model No. SCP 100
USA, 1982

This popcorn maker was part of Black & Decker's line of Spacemaker products that could be mounted on the underside of a shelf or cabinet. To operate, kernels are poured into a removable drawer and are heated in a transparent popping chamber—offering users the delight of watching the snack spring to life. Ready-to-eat popcorn falls out the chute and into a bowl below. With its dark-brown color palette, the popcorn maker looks more like a piece of AV equipment than a kitchen device, something that the packaging capitalizes on by showing the popcorn maker fixed to a media console beside the television.

The Looking Glass

General Electric | Model No. IM-5
Hong Kong, 1978

This tortoise-shell patterned electric mirror packs all the glamor of a backstage dressing room into one tidy tabletop design. The mirror is framed by a dark-brown translucent plastic with eye-catching light bars on either side, evoking dressing table lighting. The light itself is activated by a discreet on-off toggle just below the mirror, which swivels on an axis between "regular" and "magnified" views. The product's plastic shell feels sturdy and was manufactured in Hong Kong.

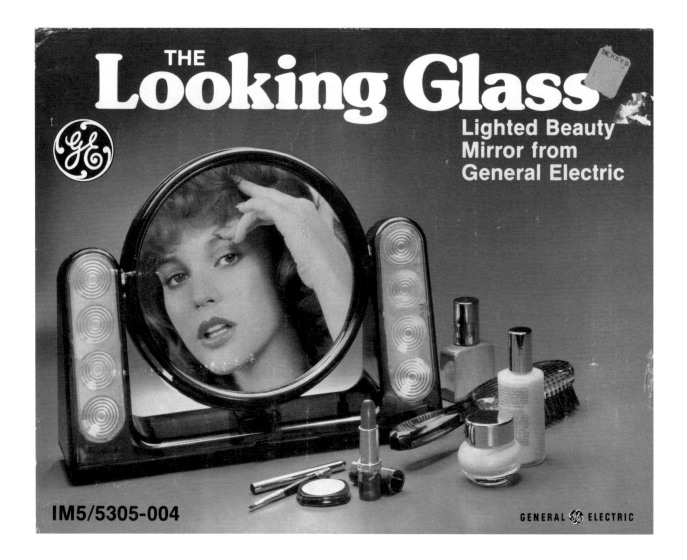

The Looking Glass | 173

Protector

Braun | Model No. PG E 1200
Germany, 1978

The Protector offered a patented heat-protection control. Users could customize the power and temperature based on their needs; indeed, one of the configurations was best suited for perms, which were sported by many in the 1970s. The Protector is one of Braun's last models with the air inlet on the sides. With subsequent models, it was moved to a single area in the back, which turned out to be a major safety improvement. The unusually short barrel almost gives it the appearance of missing a piece, but this was a feature of the design, rather than an omission.

Braun Protector®
electronic sensor hairdryer 1200 Watt

Braun Protector®
electronic sensor hairdryer 1200 Watt

BRAUN

Sanft, sogar für das empfindlichste Haar.
Gentle, even for the most sensitive hair.
Doux, protège même les cheveux fragiles.
Delicato, anche con i capelli più sensibili.
Zacht, zelfs voor het meest gevoelige haar.
Skonsam även mot känsligt hår.

Kraftvoll, für schnellstes Trocknen.
Powerful, for fastest drying.
Puissant pour un sechage plus rapide.
Potente, per un'asciugatura più rapida.
Krachtig, voor snel drogen.
Effektiv för snabb torkning.

Natürlich, für lockige Frisuren.
Natural, for curly (permed) styles.
Naturel pour cheveux bouclés ou avec permanente.
Naturale, per la moda di oggi.
Natuurlijk, voor krullen (permanent).
Naturlig för lockiga (permanentade) frisyrer.

Individuell, für Ihre Lieblingsfrisuren.
Individual, for your favourite styles.
Idéal pour réussir toutes les coiffures à la mode.
Individuale, personalizza la vostra acconciatura.
Individueel, voor uw favoriete kapsel.
Individual för Dina favorit frisyrer.

Automatik Toaster

Moulinex | Model No. 512
France, 1978

By the late 1970s, toasters became wider, offering greater compatibility for different types of bread. This Moulinex toaster, which promised on the packaging that it can toast "even French stick loaves" is a notable example of this. Its boxy, geometric design included an atypical push-down operation bar, with integrated heating setting. Users could determine how toasted they wanted their bread, and the device would function accordingly. The white shell indicated a shift in design for Moulinex (and kitchen products in general), away from the ubiquitous orange to a broader, lighter palette. Its rectangular form echoes Braun's HT2 toaster, released in 1962.

Folienschweissgerät

AEG | Model No. FSG 102
Germany, 1979

The Folienschweissgerät is a device for dispensing and sealing plastic food storage bags. The bags, which were sold in rolls, could be stored inside, and sealed and trimmed in one step. A device like this one, geared towards batch cooking and keeping food fresh longer, was especially useful at a time when women were increasingly entering the workforce but still responsible for the lion's share of domestic work. Plastic bags for food storage were in their nascence at this time; the Ziploc bag with the plastic zipper was released in the United States only a decade prior to this machine.

Folienschweissgerät

Cafetière Expresso

Calor | Model No. 22.81
France, 1976

Encased in a yellow metal shell instead of the more common plastic, the Cafetière Expresso is an easy-to-use machine that functions more like a percolator than a traditional espresso maker and comes with a small carafe that holds up to four cups.

A special attachment allows the user to brew two small cups simultaneously. The "x" in "Expresso" was a deliberate misspelling, most likely to communicate speed and efficiency. This product was sold by both Calor and Tefal.

Water Cleaner

Philips | Model No. HR 7470
Japan, 1978

This extraordinary machine was only available on the Japanese market. The water-filtration device designed by Alister Jack comes with a durable-plastic shell and an on-off switch in its base. To remove impurities from the water, it uses a charcoal filter, accommodating up to one liter of liquid. The removable cover also serves as a jug; water could be transferred from the filter to the jug using a handy, inbuilt water pump. Interestingly, the packaging sells this as more than a filter for drinking water, suggesting that it can improve the taste of soups and the aromas of teas and coffees.

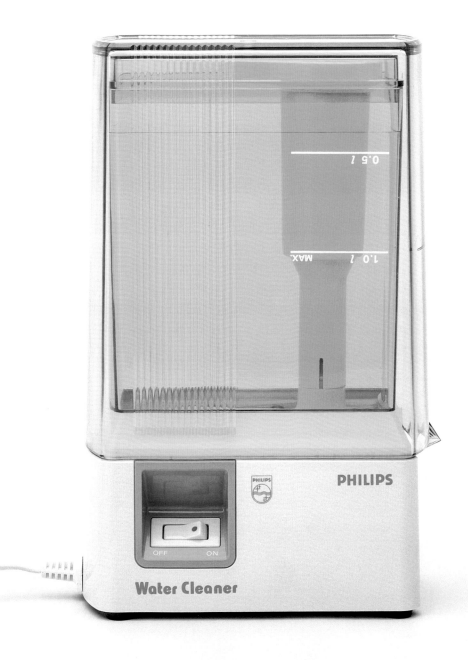

Mixer

Philips | Model No. HR 1187
Netherlands, 1978

A highlight of mid-1970s mixers, the HR 1187 was so popular that it remained on shelves until 1985. This stand mixer designed by Peter Stut featured a compact design with the same white plastic and orange accents that became the brand's design language for small domestic appliances. For this easy-to-use and easy-to-clean device, the most intriguing function is its spinning bowl. Made from a matte transparent plastic, it stands on a free-rotating disc and is spun by a little mechanism in the body of the mixer to help mix dough. The mixer could be removed from the base and used as a hand mixer as well.

Joghurt-Bereiter

Rowenta | Model No. KG-76
Germany, 1976

With its round yellow base and brown, transparent cover, this device might look like an egg boiler but its actual purpose is to heat seven small bottles of milk at a slow tempo for up to six hours. The result? Homemade yogurt, made with minimum effort. This extremely simple device didn't even come with a power switch, just an operation light that sprang on when the device was plugged in. This product was sold in several different colors and was sold by other brands with minor design tweaks.

Yogurt Maker

Toshiba | Model No. TYM-100
Japan, 1983

Despite its nondescript appearance, this yogurt maker has unusual features. Unlike comparable European devices, often composed of several glass jars for individual portions, this example by Toshiba is just a single container in an all-plastic design. The logic was that fewer components meant less cleaning. The one-liter container comes with its own screw-on lid and can be stored in the fridge. Its hard plastic shell and color are typical of mid-1980s design.

Novodent Pulsar

Krups | Model No. 353
Switzerland, 1982

The Novodent Pulsar oral irrigator was sold alongside the Novodent Vario electric toothbrush. It was designed according to the Krups design language, with soft rounded edges, and its base was available in white or green with a clear transparent water tank. Previously, water would have been stored in opaque plastic tanks, obscuring chalky residue. The Novodent line would be Krups's first and last collection of dental products. They concluded early on that there wasn't sufficient medical evidence for oral irrigators as an alternative to flossing, compounded by the fact that Braun was already dominating the market.

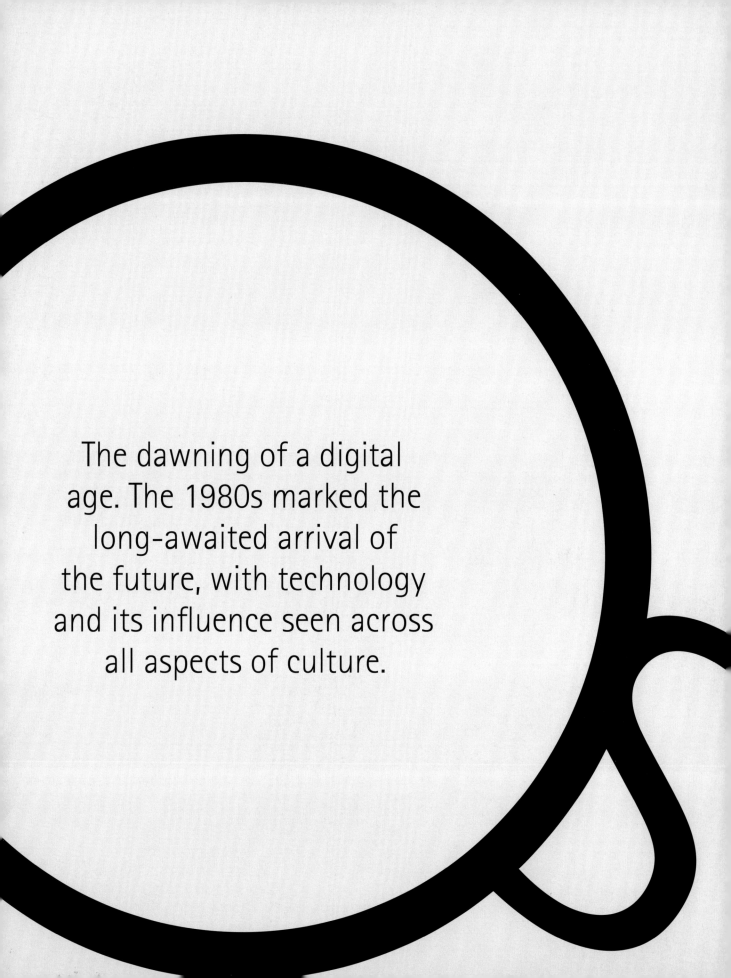

The dawning of a digital age. The 1980s marked the long-awaited arrival of the future, with technology and its influence seen across all aspects of culture.

Technology was starting to play an important role in people's lives. In 1982, *Time* magazine gave its Man of the Year award to the "personal computer."

n 1982, *Time* magazine gave its Man of the Year award to the "personal computer." While that's not a term often heard anymore, it says a lot about the 1980s. As its title suggests, the award was normally a men's club, the only deviation from this prior to 1999—after which the trophy became "Person of the Year"—being four "Woman of the Year" honors. As such, the 1982 trophy was unique, and made clear the extent to which technology was transforming society.

Despite heralding the dawn of the Information Age, in which society shifted from reliance on industry to technology, the decade began on a somber note. John Lennon's murder on December 8, 1980 shocked the world. Later, in 1981, the HIV/AIDS epidemic began, and rising oil prices continued to rock the West. Elsewhere, while women had begun to enter executive positions, in 1986 *The Wall Street Journal* coined the term "glass ceiling" to describe the invisible workplace barriers women continued to face. Then as now, gender equality was a battle still being fought.

By 1983, many Western economies had rebounded and were enjoying sustained growth. Inflation rates stayed low for the remainder of the decade, driving middle-class wealth. Families and young professionals were moving out of the city to the suburbs, where they were buying big houses with spacious kitchens and living rooms, perfectly manicured lawns, double garages, and curb appeal aplenty. These newly populated areas were often filled with people who had migrated from industrial cities, exasperated by overcrowding, pollution, and crime.

Inside the home, change was also afoot, not only in terms of chintzy redecoration but in family roles and dynamics too. It was now more common for women to work full-time jobs, although, as in the '70s, this did not mean they were relieved from their domestic duties as housewives and homemakers. The invention of the countertop microwave sped things up a little, with cookbooks in the '80s detailing methods for cooking entire meals inside it. *The Microwave Cookbook* (1984) by British chef Carol Bowen, even featured a recipe for French classic Lobster Thermidor. Placing an entire lobster inside a microwave may seem unthinkable now, but the '80s were all about having everything as quickly as possible.

This lust for convenience crept out of the kitchen and into the living room, which was typically decorated in pastel colors like lavender and peach, with mottled prints for upholstery, glass surfaces, and linoleum floors. Remote controls for the TV meant you no longer had to get up to change shows—and when it came to channels, the choices were ever-expanding. MTV launched in 1981, bringing with it a new way to enjoy music: the expressive,

Families and young professionals were moving out of the city to the suburbs, where they were buying big houses with spacious kitchens and living rooms, perfectly manicured lawns, two-car garages, and curb appeal aplenty.

and often bonkers, music video. Michael Jackson's "Thriller" (1982) is credited with transforming the music video into a serious art form, and many elements from it have had a lasting impact on popular culture, such as its iconic zombie dance and Jackson's red leather jacket. When not watching MTV, young people were living through the golden age of teenage cinema, with cult classics such as *The Breakfast Club* (1985), *Sixteen Candles* (1984), and *Ferris Bueller's Day Off* (1986) proving hugely popular.

During the decade, the kitchen also changed. No longer a function-over-form, cooking-only quarters, it became the central hub of the home, with island benches and breakfast bars making eating a casual occasion. Gadgets were everywhere, with machines designed for very specific purposes coming onto the market. Black & Decker's Popcorn Center (1982) puffed kernels ready for a night in front of the TV, Calor's La Chocolatière (1985) created velvety smooth hot chocolates on demand, and to provide a little extra ambiance, Aromance's Aroma Disc Player (1983) diffused fragrance from a set of "records."

To the relief of everyone's dentist, in the late '70s and '80s home-appliance manufacturers began to focus on oral hygiene products. In 1980, Rowenta's Dentasonic came along, making it easier for dentures to be scrubbed. A year later, AEG released its Dentalux AE, a cordless electric toothbrush that paved the way for similar models. Its stripped-back white plastic casing signaled a move towards a more restrained approach, a gesture that would become typical of '80s design. The psychedelic hues of the '60s and '70s took a backseat, and appliances were given serious aesthetic treatment. The bright oranges seen so commonly on blow-dryers, coffee makers, coffee grinders, and mixers in the '70s, were replaced by whites and grays to match white-tiled kitchen tops and clean surfaces.

As colors changed, materials did too. The production techniques associated with manufacturing plastics improved to allow for shinier, glossier surfaces, as well as experimental shapes and forms thanks to more malleable materials. "After the colorful '70s, most brands had to change their colors to white to indicate clearly that it was a newer device. It was also a tool for consumers to leave the 1970s behind," says interaction designer Jaro Gielens, whose collection is featured throughout this book. In the '80s, the image (or memory) of the '70s was often

The invention of the countertop microwave sped things up in the kitchen, with cookbooks in the '80s detailing methods for cooking entire meals in it.

No longer a function-over-form, cooking-only quarters, kitchens became the central hub of the home.

connected with smudgy or impure times, and in settings like kitchens and bathrooms there was a demand for clean looks."

This aesthetic was not quite mirrored in fashion and clothing, as in this respect, the '80s was a maximalist decade. Hair was big, and shoulder pads were even bigger. Neon-pink eyeshadow clashed with neon-lime legwarmers. Sports brands, skiwear, baggy jeans, and Ray-Ban sunglasses provided the perfect rave outfit, whereas for businesswomen, sharp-shouldered power dressing was *de rigueur*.

In design, the Memphis movement combined eye-popping patterns with haphazard color palettes and geometric shapes in a bid to poke fun at the function-first style that came before it. New styles in music were emerging, too. Grunge, famously born in Seattle, saw the formation of bands such as Nirvana, Soundgarden, and Alice in Chains. Elsewhere, electronic synthesizers, drum machines, and samplers were being used by musicians in Detroit and Chicago to create techno and house respectively.

For those born after the 1980s, it is hard to comprehend the effect technology had on day-to-day life. On the other hand, we seem to have gone back to basics in a lot of ways. Style has, for the most part, become more subdued, our surroundings are softer, and our lobsters are cooked in ovens. But given that the 1990s gave way to the notion of planned obsolescence—an idea that seems antithetical to current eco-conscious consumer demands—perhaps there is a thing or two that can be learned from looking back.

TS10 Aroma Super Luxe

Krups | Model No. 163
Germany, 1979

For the pinnacle of late 1970s coffee makers, look no further than the Aroma Super Luxe. It features an easy to use high-tech control panel with several settings. The device also introduced a now-classic turquoise VFD digital display, which uses the same technology as alarm clocks. With its elegant form and futuristic features, consumers had no qualms with the upper-end price point. This was something the brand learned the hard way when it released the Cafethek Luxe in 1981, which was intended to build on the success of the Aroma Super Luxe. The result: its many bells and whistles confused consumers—and no longer justified the price.

Multipractic Plus

Braun| Model No. UK 1
Germany, 1983

This style of food processor originated in Japan in the 1970s, and though Braun's competitors had been selling similar products for a number of years already, Braun released its first food processor relatively late with this Multipractic Plus. Better late than never, it proved to be a machine with many talents. The kit is composed of four different slicing discs and a blade, meaning the processor can whip, mix, beat, slice, and shred, depending on how it's set up. This model offers a safety feature—something that wasn't always a given—in that the mixing container had to be locked in to activate the machine.

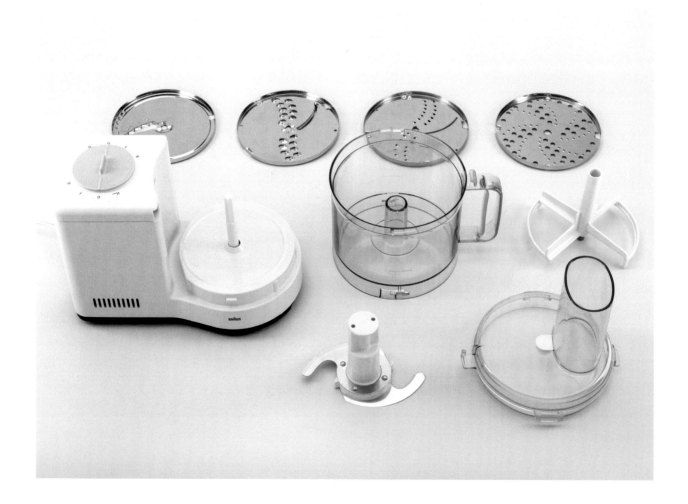

Softstyler

Braun | Model No. PG S 1000
Germany, 1982

Without the diffuser, the Softstyler would be just another pistol-shaped blow-dryer, of which Braun had already released many. But as soon as this big, round attachment is put in place, it becomes an entirely different product. The air ventilation holes on the front resemble those on the back. The megaphone-like shape becomes an immediate shorthand for big 1980s hair. Another spectacular element of this piece is the shade of green, an excellent hue of '80s avocado, which is the only color in which the product was available.

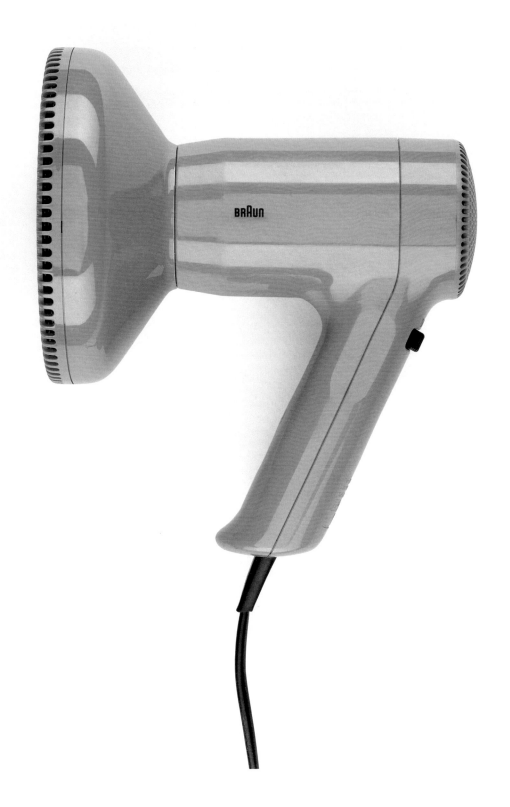

Stabmixer Vario

Braun | Model No. MR 6
Germany, 1981

The Stabmixer Vario appeared nearly 31 years after the invention of the immersion blender by Roger Perrinjaquet, the Swiss founder of appliance company Bamix. Designed by Ludwig Littmann, its smooth white shell has an ergonomic indent for controlled grip and is completed by a stainless-steel shaft and polished basket. It comes with multiple speed settings, which can be controlled by a dial at the top and a bright red power switch. Pictured is the complete deluxe set, which came with two pitchers, a sieve, a mixer attachment, and even a wall mount.

3 Mix 4004

Krups | Model No. 727
Germany, 1982

Despite its prosaic name, the 3 Mix 4004 became a best-selling product. The simple, straightforward design by Rudolf Maas was the latest, modernized product from Krups's family of 3 Mix mixers, which first appeared on the market in 1959. As with prior models, additional attachments were sold separately, including the unusual "schnitzelwerk" intended for shredding. It's likely that many of these mixers are still in use in German households today, as these products are virtually unbreakable.

Ⓓ Ausbaufähig zum vielseitigen Küchensystem.
🆎 Can be extended by various accessories into a multipurpose kitchen system. (✱ not available in the U.K.)
Ⓕ Peut être complété en un système robot de cuisine à usage multiples.
Ⓝ Geschikt om te worden uitgebreid tot een veelzijdig keukensysteem.
Ⓔ Ampliable a sistema robot de cocina.
Ⓢ Kan byggas ut till ett mångsidigt kökssystem.

Ⓓ Starker 170-Watt-Motor. Momentschalter für Kurzbetrieb. 3 Geschwindigkeitsstufen.
🆎 Powerful 170 watt motor. Instant button for short operation. 3 speeds.
Ⓕ Puissant moteur de 170 Watt, commutateur instantané pour un bref fonctionnement. 3 vitesses.
Ⓝ Sterke 170-watt-motor. Momentschakelaar voor kort gebruik. 3 snelheden.
Ⓔ Motor de 170 W de potencia. Interruptor momentáneo para uso breve y 3 velocidades.
Ⓢ Kraftig motor på 170 W. Momentomkopplare för kortvarig körning. 3 hastighetssteg.

Ⓓ Platzsparende Wandhalterung.
🆎 Space-saving wall-bracket.
Ⓕ Support mural peu encombrant.
Ⓝ Plaatsbesparende wandhouder.
Ⓔ Sujección de pared para ahorrar espacio.
Ⓢ Utrymmesbesparande väggfäste.

Ⓓ Bedienungsfreundlich durch Spiralkabel.
🆎 Spiral cable for easy use.
Ⓕ Confort de manipulation par cordon en spirale.
Ⓝ Gemakkelijk te bedienen door het spiraalvormige snoer.
Ⓔ Manejo más cómodo con el cable espiral.
Ⓢ Lättskött tack vare spiralsladden.

La Pasta Machine

Moulinex | Model No. 717
France, 1983

Bring a little taste of *la dolce vita* right into your 1980s kitchen with this deluxe pasta maker from Moulinex. The serif typeface gives the impression it's chiseled into an ancient Roman artifact, a nice visual detail. Other features are less forgiving: the product comes with a heavy motor and relies on many additional parts and pasta discs. It's complex but high quality and very functional. This particular example is an American model sold in cooperation with Regal under the Regal Moulinex brand.

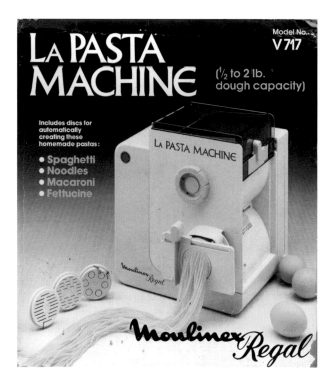

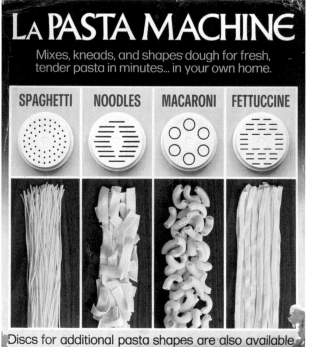

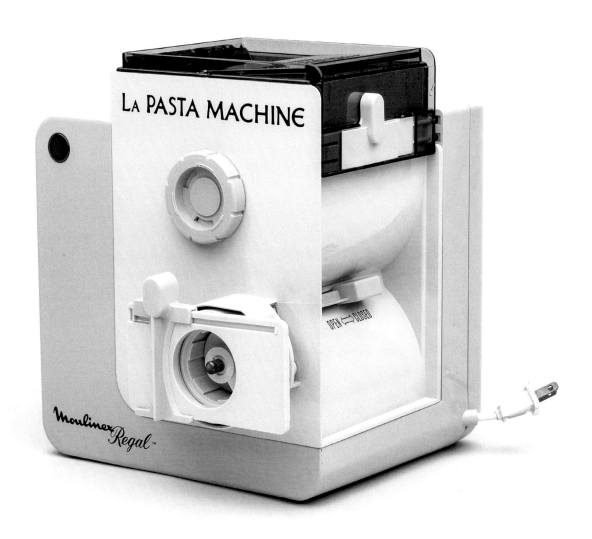

Ice Crusher

Toshiba | Model No. KC-55A
Japan, 1984

Crushing ice in a blender tends to cause damage and dull its blades, a problem that this machine alleviates. The futuristic design of this Japanese ice crusher resembles the structure of some of the larger coffee grinders in this collection.

The crushing mechanism is visible through the transparent cover, allowing users to choose how finely they want their ice ground. Owing to its compact size and the fact that it's battery operated, the KC-55A is easily portable.

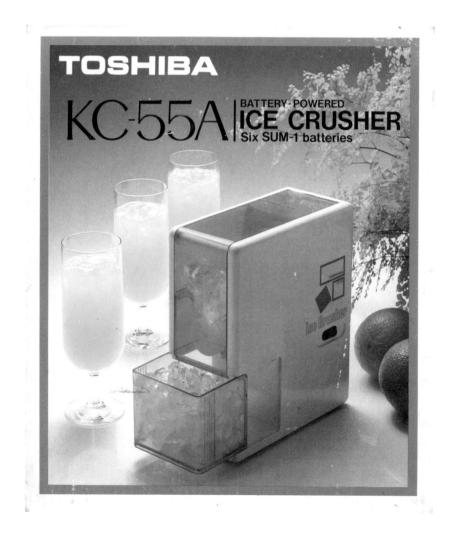

Travel Iron

Philips | Model No. TI 6500
Taiwan, 1982

For those who need their shirts pressed on the go, Philips released this angular travel iron designed by Ron Muller in 1982. For easy portability, it's smaller than a regular iron, and its handle can be folded down into a flat, compact shape. Although the body of the device appears to be made of metal, it's actually a robust plastic, while the base is made of aluminum. It also comes with a small, removable water container for spray ironing and steam. The Travel Iron features a similar design approach and color scheme to the handheld Portable Fabric Steamer, which was released in the same year.

Fashion

Rowenta | Model No. DA-54
Germany, 1985

Rowenta had already been in the business of producing irons and travel irons for 66 years when they released this playful product in 1985, which was sold in a clear display box. The light handle could fold over to turn the iron into a flat, easy-to-pack block, and it also had a steam function with a water indicator window. The Fashion came in a fresh palette of white with sunny yellow accents–the yellow color being what prompted the name "Fashion." Rowenta released a very similar product with blue details just a few years prior, and a pink version was released in the late 1980s.

Mr. Instant

Melitta | Model No. MI-500
Japan, 1981

This Japanese-manufactured product was part of the German brand Melitta's strategy to breach the Asian market, which it attempted with four products in the "Mr." series, including a food processor and ice cream maker. Its design, which uses different plastic finishes to give the appearance that it's made of metal, doesn't immediately reveal its function. The Mr. Instant could make multiple types of hot beverage, and featured a handle for easy portability and a little door to insert and remove mugs. It was sold with a jug and a set of Melitta paper filters.

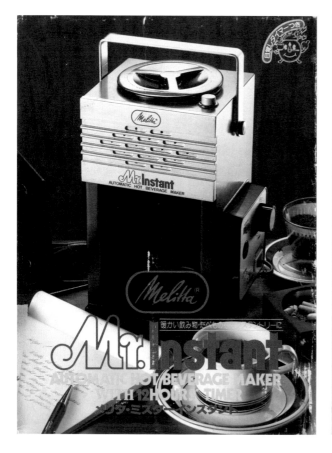

Mr. Instant | 217

Air Cleaner

Philips | Model No. HR 4371
Japan, 1980

This handsome air cleaner was one of several mini filtration devices released by Philips, all of which were manufactured in Japan. Designed by Ron Muller, the product was typically available in white but this upgraded version came with an imitation wood and hard black plastic shell. The device draws-in air through its electrostatically charged filtration system, attracting dirt and other pollution particles. From there, it recirculates the filtered air. A special clear plastic shell houses the filtration discs for the user to see when it needs replacing.

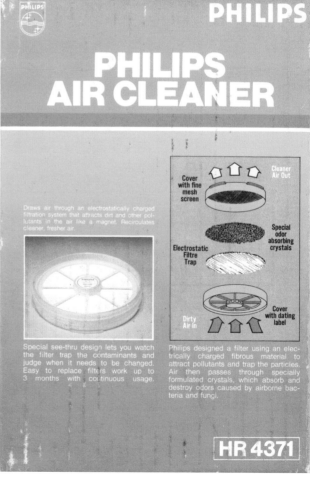

Aroma Disc Player

Aromance | Model No. 4300
Hong Kong, 1983

Many products in this book have stood the test of time, but the Aroma Disc Player, a home fragrance diffuser where scented "fragrance records" are inserted like floppy disks, is not among them. The player heats the disk and the resulting scent is dispersed with a small inbuilt fan. Fragrance records were sold separately and included scents like "A Dozen Roses," "Seduction," and "Movie Time" (which smelled like buttered popcorn.) The product enjoyed decent sales figures after release and was followed up with the Aromance 2100, which let the user control the intensity.

La Chocolatière

Calor | Model No. 2350.02
France, 1985

Calor's 1985 hot chocolate maker was a very sophisticated design for a machine that really only has a single specific function. The base is wrapped in a thin layer of cork, which conceals the heating mechanism, though later models ditched the cork and were sold with a simple white plastic base. In the vessel, milk is heated up along with chocolate chunks, and the result, according to the packaging, is smooth, old-fashioned chocolate. This product came complete with a recipe book to inspire users with chocolatey recipes.

Simple à utiliser: le bol de préparation est déjà gradué.

Un chocolat mousseux et onctueux obtenu grâce à un système de jets de vapeur.

Entretien pratique: bol de préparation séparé de la base.

S'utilise avec tous les types de lait et de chocolat.

Parfaite sécurité: témoin lumineux de branchement.

Livrée avec un livre de recettes gourmandes.

pour faire un authentique chocolat à l'ancienne, onctueux et mousseux à partir de chocolat en morceaux.

ne nécessite aucune surveillance: le lait ne peut pas déborder.

un appareil astucieux et plein d'idées: livré avec un livre de recettes complet.

Bimbo

Rowenta | Model No. KG-39
Germany, 1983

Rowenta's early-1980s solution to heating up milk, formula, and baby food, took the form of a plastic elephant. Glass bottles and jars are simply placed into its cylindrical compartment, where the temperature is controlled by a circular dial on the elephant's trunk. This charming product is often credited to Luigi Colani, the German industrial designer who created everything from trucks to coffee cups. It's more likely, however, that it is an adaptation of the Drumbo piggy bank, designed by Bernd Diefenbach, to which this product bears an unquestionable resemblance.

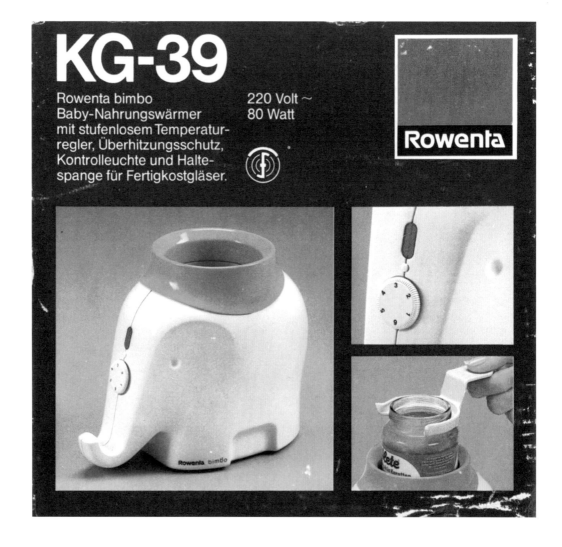

HotTopper

Presto | Model No. 3000
Hong Kong, 1986

Perhaps one of the best qualities of kitchen devices from the 1970s and '80s is the way they solve problems we didn't even know we had. The HotTopper is a good example of this. Its canister could heat and melt ingredients like chocolate, butter, syrup, and sauces, and the user could fasten an attachment for spraying, brushing, or pouring the topping on their meal. Ideal for spraying butter on popcorn, pouring syrup on pancakes, or even brushing barbecue sauce on meat. The product was a success and remained on the market for more than a decade. A later microwave version with a more modern design was released a few years later.

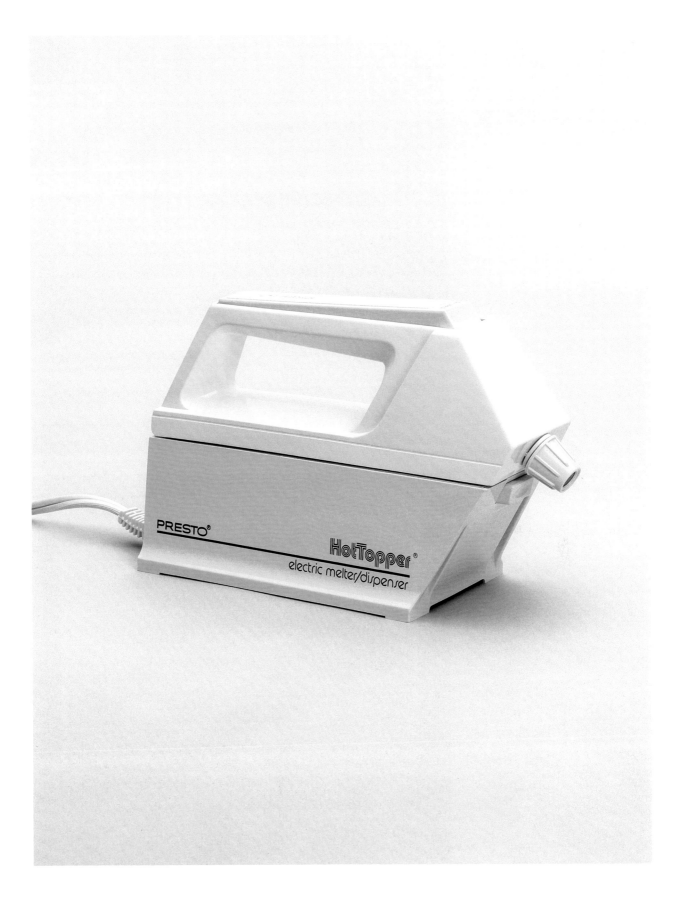

"No need to clean after each use"—seems that Presto wanted to sell more products.

- **Automatically melts butter or margarine...heats syrup and other great toppings, too.**
- **Easy, thumb-action pump sprays, streams or brushes delicious hot toppings on your favorite foods.**
- **Cord removes for extra convenience. Use it right at the table, counter or grill as a cordless appliance.**
- **No need to clean after each use. Leave unused toppings right in the unit and store in the refrigerator to use again and again.**
- **Completely immersible for quick and easy cleaning.**

Sprays evenly for perfect hot-buttered popcorn.

ot butter and syrup on
s, waffles.

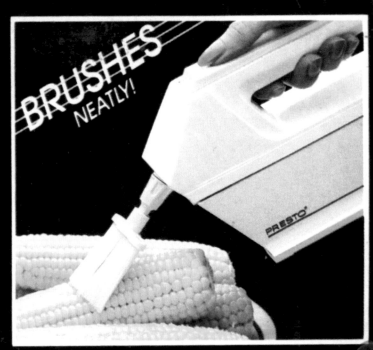

Brushes hot butter for delicious corn-on-the-cob.

butter for super
aked potatoes.

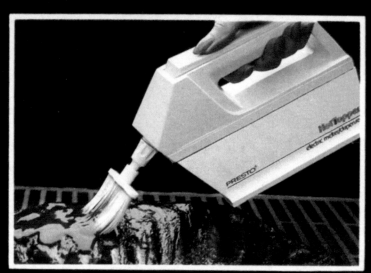

Brushes for easy barbecuing without all the mess.

Chefette de-Luxe

Kenwood | Model No. A 380
United Kingdom, 1981

This British-made product, which wouldn't look out of place preinstalled in a Barbican apartment, is a powerhouse among cake mixers. It's a two-in-one machine: one side a stand-mixer for beating eggs and batter in a transparent plastic bowl, the other equipped with a half-liter blender. The design exemplifies the modernist aesthetics of the 1980s, especially its octagonal shape and color palette, which the packaging quaintly refers to as "country beige and brown." Conveniently, the device could be removed from its housing and used as a hand mixer.

Chefette de-Luxe | 231

Aroma Art

Melitta | Model No. 9040
Germany, 1984

At first glance, it isn't immediately clear if this object belongs in the laboratory of a mad scientist or in the kitchen of a German family. Its shape suggests a very scientific approach to making the perfect home brew, evidenced also by the "unique thermo-tech design" advertised on the packaging. The composition is comprised of three distinct bowl-like elements, and in the water tank, the red and white water level indicator bobs like a fishing float. The modernist design, which was also available in red and pink, is by the German industrial design studio OCO Design.

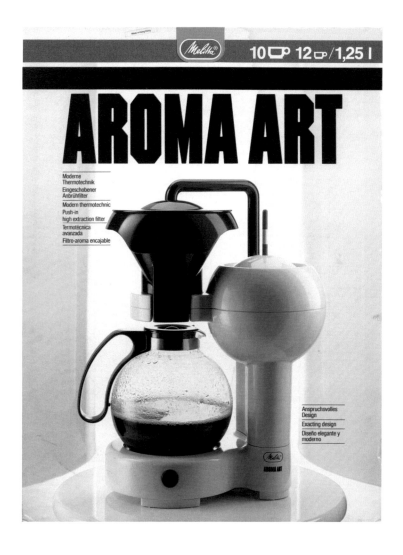

Turbo Jet 1205

Philips | Model No. HP 4125
Germany, 1977

This was considered a top-of-the-line blow-dryer which, as its aviation-inspired name suggests, offered its users maximum power. It was a stylishly designed piece by Peter Nagelkerke, coming in an unusual mustard yellow plastic with a circular black accent encasing the air inlet. The device also shares a few similarities to the Italian Superphön, which was developed by Philips outside Holland two years prior. The Turbo Jet 1205 was sold with a black styling nozzle and a stand, which propped it up, and sent a clear message that this product isn't to be folded up and stored away, but showcased for all to see.

Many of Philips packaging photos in the 1970s were shot by well-known advertising photographer Christopher Joyce in a studio in London, UK. Packaging design by Henk Jan Drenthen.

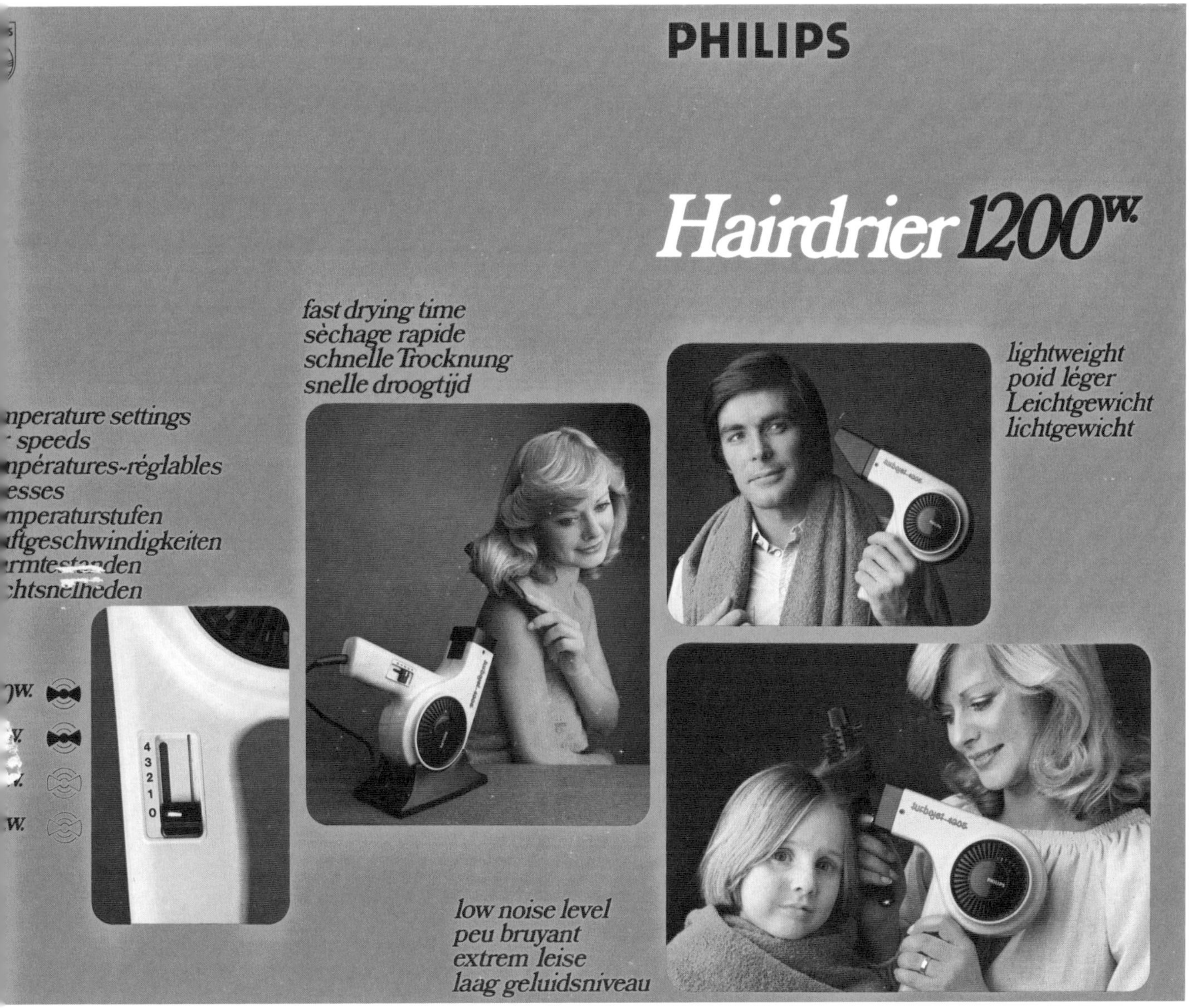

PHILIPS

Hairdrier 1200ʷ·

fast drying time
sèchage rapide
schnelle Trocknung
snelle droogtijd

lightweight
poid léger
Leichtgewicht
lichtgewicht

...nperature settings
 speeds
...npératures-réglables
...esses
...nperaturstufen
...ftgeschwindigkeiten
...rmtestanden
...htsnelheden

low noise level
peu bruyant
extrem leise
laag geluidsniveau

Turbo Jet 1205 | 237

Plus 400

Philips | Model No. HR 2986
Netherlands, 1980

In 1977, Philips released its first kitchen machine, the HR 2970, which was in development for three years. It was worth the wait, becoming one of the most popular products from their Small Domestic Appliances (SDA) division, selling in high numbers across Europe. The product pictured here, the Plus 400, is its follow-up from 1980, which only had a few small tweaks. Designed by Hans Elkerbout, its arm was made tiltable, making it easier for users to change attachments and fasten the bowl. The attachments for the original machine were also compatible with this update, which was available well into the mid-1990s.

BOX 2

Philips | Model No. HR 2010
Netherlands, 1983

The BOX 2 can be likened to a transformer among kitchen devices. With different fixtures and fittings, it can be folded out into a stand mixer or reconfigured into a hand mixer. These different functions are all based around one central component housing the motor. This was a classic feature of the BOX series, designed by Hans Elkerbout, which offered several different product variations, all focused around the theme of easy reconfiguration. For instance, the BOX 7 was a stand mixer, mincer, peeler, and coffee grinder in one and complete with a custom storage cabinet for all parts. Unfortunately, the line didn't achieve the commercial success that its makers envisioned and was discontinued after just one year.

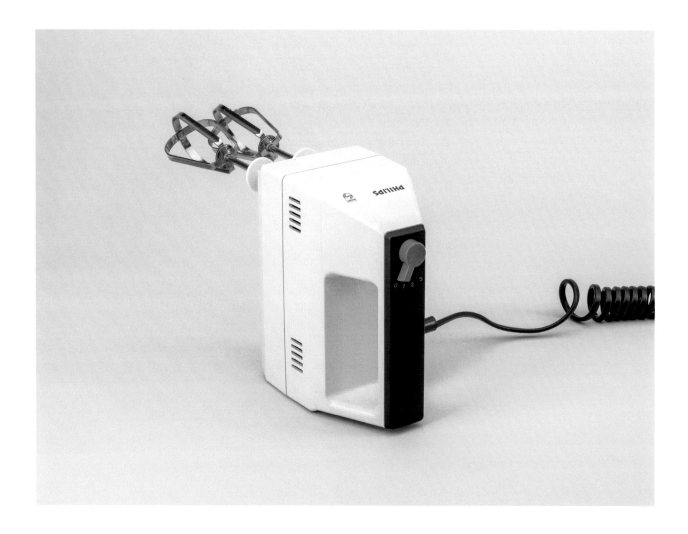

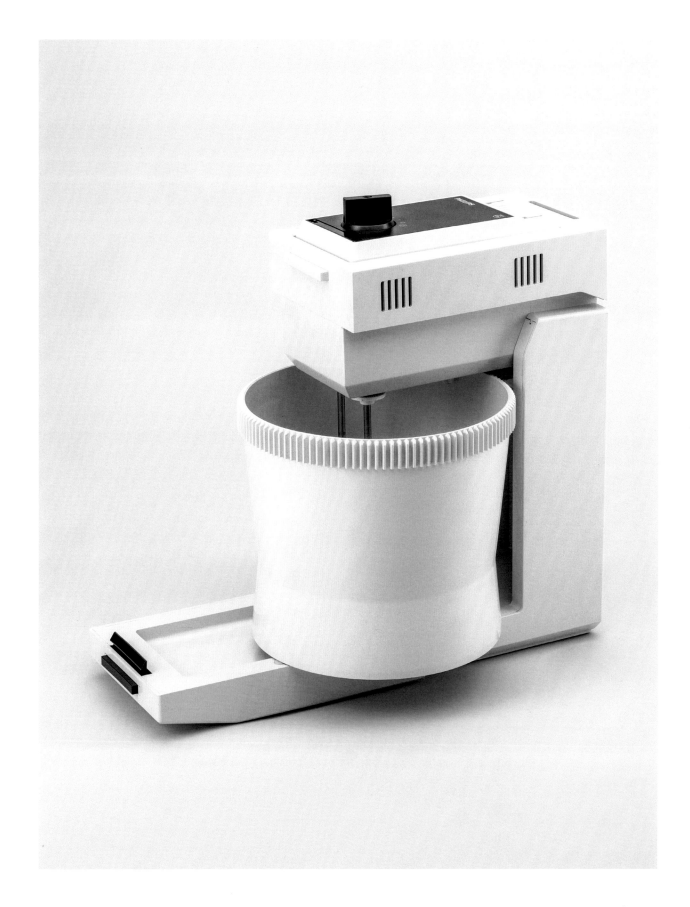

This design is exactly in the transition from the black packaging of the 1970s to the white packaging of the 1980s with a very prominent Philips logo. Packaging design by Karl Knoflach.

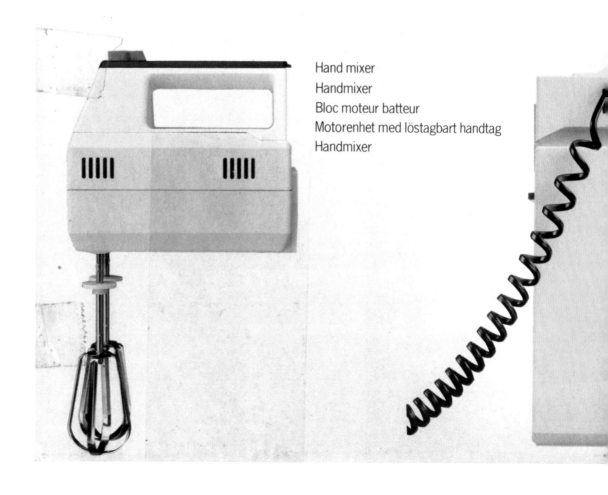

Hand mixer
Handmixer
Bloc moteur batteur
Motorenhet med löstagbart handtag
Handmixer

PHILIPS
Fabriqué en Hollande

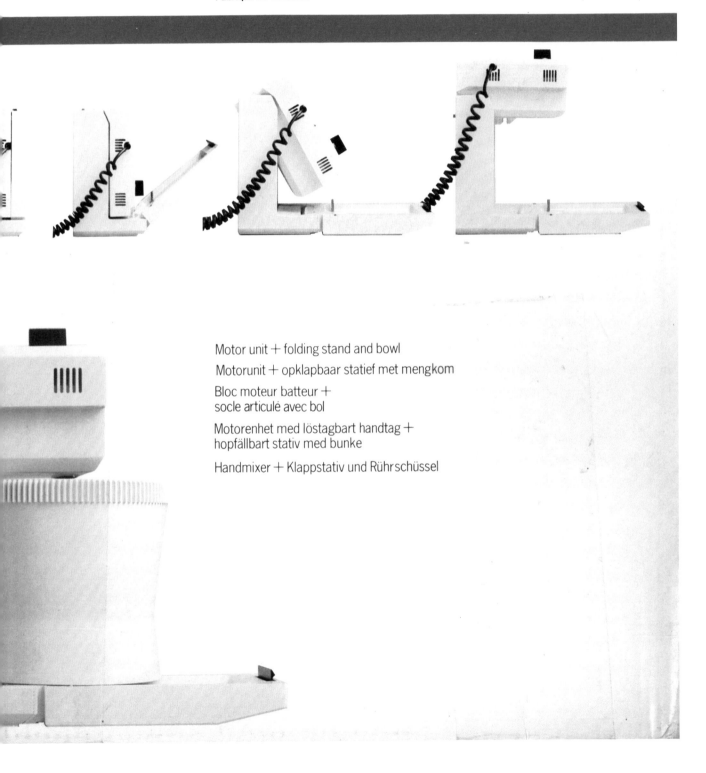

Motor unit + folding stand and bowl
Motorunit + opklapbaar statief met mengkom
Bloc moteur batteur + socle articulé avec bol
Motorenhet med löstagbart handtag + hopfällbart stativ med bunke
Handmixer + Klappstativ und Rührschüssel

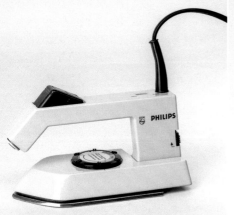

Light bulb moments. From domestic appliances to digital technology, the Dutch company's history is one of cutting-edge innovation.

Philips

The Philips stand at the Salon de la TSF, held at the Palais de la Découverte in Paris, 1935.

Advertisements for the Philips illuminated light bulb, from 1925, and the brand's cordless shaver, from 1967.

The story of Dutch company Philips is one of innovation, reinvention, and many ingenious light-bulb moments. Inspired by Thomas Edison's first incandescent light bulb, Gerard Philips, then a young science and engineering enthusiast, asked his father to purchase a small plot of land in Eindhoven, where he could set up his own light bulb factory. The factory opened in 1891, and Philips & Co. was born.

The company grew swiftly, aided by advanced industrialization and assisted by a historic contract with Tsar Nicholas II who ordered a large number of light bulbs for his Winter Palace in Saint Petersburg in 1912. In 1914, Philips launched its renowned NatLab, a groundbreaking physics laboratory dedicated to the research and development of new product ranges.

By the 1920s, the company had begun making radio components and soon started placing a stronger emphasis on design. This was propelled by its close ties to the transmitter manufacturer Nederlandsche Seintoestellen Fabriek ("Dutch signal factory"), an enterprise Philips co-founded in 1918 with Marconi UK and Radio Holland. Through this, it began producing its own radios, an endeavor that was propelled in 1924 by the appointment of designer Louis Kalff, who oversaw the company's inimitable advertising and branding.

In 1939, the company launched its revolutionary rotary electric razor, the Philishave, a slick, ergonomic device that proved an immediate and enduring success. Today, Philips claims that since the product hit the shelves, an average of 700 Philishaves have been sold every hour.

After the Second World War, the company sustained its inventive streak, embracing the rise of television and introducing its own TV set in 1949. By 1954, it had established its own design group, which began releasing a steady flow of stylish lamps, razors, radios, and gramophones, among other products.

In 1960, the brand had formed an industrial design office and, in reaction to the decade's pop-culture explosion and increased consumer spending power, set out to make its cutting-edge technologies even more accessible. It developed the first compact cassette audio player in 1963, setting a new standard for tape recording thereafter. (Keith Richards famously recorded the riff to "[I Can't Get No] Satisfaction" on his own Philips recorder.) Video players and recorders soon followed suit. Meanwhile, Philips's kitchen appliances were fast becoming household favorites, as advertisements proclaimed the brand, "the friend of the family."

In 1969, Norwegian artist and designer Knut Yran was appointed head of Philips's industrial design, tasked with implementing a consistent brand identity. This included a renewed focus on good, inexpensive design, and aesthetically pleasing packaging. His approach was evident in many of Philips's new products, including its fancifully colored, all-plastic vacuum cleaner released in 1977, and its first food processor, the HR 2670 (1977), which proved an instant classic thanks to its neat, fool-proof design.

In 1980, American industrial designer Robert Blaich succeeded Yran, and was charged with unifying the design strategy across all product sectors. An interior architect responsible for the seating at the United Nations's New York headquarters, Blaich believed Philips's products had to speak to the ethos of the company in both appearance and function, placing a strong emphasis on ergonomics, safety, utility, and efficiency.

Philips continued to blaze trails in emerging consumer digital technologies throughout the decade, devising optical telecommunication systems and releasing the first compact disk player in 1982, an enterprise it had begun with Sony in 1969. In the '80s, the brand also remained a leading competitor in the realm of home appliances and personal care.

In the decades that have followed, the vast corporation has continued to expand, invent, and evolve, shifting its focus to healthcare and personal care products in 2013.

Cafe Gourmet

Philips | Model No. HD 5560
Portugal, 1988

This trailblazing machine had its own optimized method of making coffee. It's simple: it heated the water to 93° Celsius (199° Fahrenheit), which was accepted as the ideal temperature to make the perfect brew. The tall design is atypical for the category—home coffee makers normally stand on kitchen counters below cabinets.

The Cafe Gourmet circumvented this issue through some clever marketing: the machine was portable, and even if users unplugged it and moved it to the living room, the coffee would stay warm. By breaking these rules, it afforded its designer Lou Beeren greater freedom and resulted in an iF Design Award in 1991, among others.

Café Duo

Philips | Model No. HD 5171
Netherlands, 1983

The Café Duo marks an important moment in the development of Philips's range of coffee makers. Designed in the early 1980s by Alister Jack, it served as a template for many subsequent designs, which have been tweaked and improved over time. In fact, a descendant of this machine is still on store shelves in the 2020s. Unlike filter machines with an 8–12 cup capacity, the Café Duo is a compact machine that could make just one to two cups at a time. It reflected the changing lifestyle of its customers: busy professionals who lived alone and didn't need family-size kitchen devices.

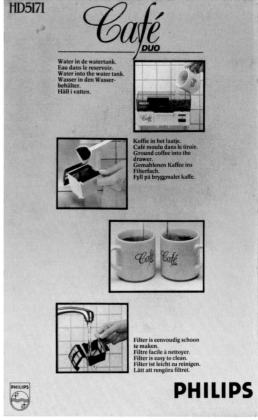

Aromaster 10 Control S

Braun | Model No. KF 90
Germany, 1986

This is the deluxe edition of a similar Aromaster coffee maker—the KF 40 (1984), which was also designed by Hartwig Kahlke—and became a blueprint for Braun coffee makers in the 1980s. Unlike other drip coffee makers, this one opted for a thermal, plastic coffee pot instead of glass. The result is a clean appearance in white plastic. The downside is that the scale is printed on the inside of the water tank only, making it difficult for the user to measure the water quantity. It also features an illuminated LCD display for its digital clock and timer function.

Aromaster

Braun | Model No. KF 43
Germany, 1990

If this 1990 Aromaster looks familiar, it's because it's been on store shelves for more than 30 years. This version takes the Aromaster KF 40 from 1984 and gives it a technical update, especially on the lid. The design of the coffee pot stops the vapor from escaping, promising a more flavorful brew. It also includes a drip-stop function, a generous 15-cup capacity, and a pivoting filter holder. On the right side, a clear plastic compartment indicates the water lever with a little red floating marble, making it slightly easier to measure water levels than its earlier iteration, the KF 90.

Ovomat Trio

Krups | Model No. 234
Germany, 1983

By 1983, Krups had already released a series of high-quality and well-designed egg boilers. With the Ovomat Trio, a downsize three-egg version of previous iterations, designer Hans-Jürgen Precht solved an existing design issue: not everyone needs six or seven eggs boiled at once. Fewer eggs also means fewer materials for production, and less space taken up in the kitchen. With a smaller footprint, it could easily slide into a narrow gap in a kitchen cabinet or stand unobtrusively against the wall. Its cord also wraps into the product's base.

Index

All product photography is by **Studio Sucrow** and packaging scans by **Jaro Gielens**, unless otherwise stated below.

All trade names, trademarks, service marks, logos, and designs belong to (or their legal successors):

AEG: **Electrolux AB**
Aromance: **Environmental Fragrance Technologies, Ltd.**
Black & Decker: **Stanley Black & Decker Inc.**
Bosch: **Robert Bosch Hausgeräte GmbH**
Braun: **BRAUN P&G/Braun Archiv Kronberg**
Calor, Krups, Moulinex, Rowenta, SEB, Tefal: **Groupe SEB Deutschland GmbH**
Emide: **EMIDE Germany GmbH**
ESGE: **Unold AG**
General Electric: **GE Appliances**
Gillette: **P&G**
Girmi: **Girmi SpA**
ITT: **Thales Group**
Melitta: **Melitta Unternehmensgruppe**
Kenwood: **De'Longhi Group**
National: **Panasonic Corporation**
Philips: **Koninklijke Philips N.V.**
Presto: **National Presto Industries, Inc.**
SHG: **SHG Vertriebsgesellschaft für Hausgeräte mbH**
Toshiba: **Toshiba Corporation**

Additional images:
Back cover, top left: Pictorial Press Ltd/Alamy; p. 5: Found Image Holdings/getty images; pp. 7, 13 top left: Archive Photos/getty images; p. 8: H. Armstrong Roberts/ClassicStock/getty images; pp. 9, 12 top right, 14 bottom left, 56: Retro AdArchives/Alamy Stock Photo; p. 10 top left: Martyn Goddard/Alamy Stock Photo; p. 10 top right: Sunbeam Vista Steam Iron Owners Manual User Guide; p. 10 middle: www.faema.com; p. 10 bottom left: Picture Post/Hulton Archive/getty images; p. 10 bottom right: George Crouter/The Denver Post/getty images; p. 11 top left: Mondadori Portfolio/getty images; p. 11 top middle: Dennis Hallinan/getty images; p. 11 top right: Thomas Drebusch; p. 11 bottom left: Koninklijke Philips N.V.; p. 12 middle: Apic/getty images; pp. 12 bottom left, 13 top middle: Suzanne FOURNIER/getty images; p. 12 bottom right: David Cooper/Toronto Star/getty images; p. 13 bottom left: Stevens/Fairfax Media/getty images; p. 13 bottom right: www.singer.com; p. 15 top middle: Terry Schmitt/UPI/Alamy Stock Photo; p. 32: H. Armstrong Roberts/ClassicStock/Alamy Stock Photo; pp. 33 left, 34: Neil Baylis/Alamy Stock Photo; p. 33 right: Patti McConville/Alamy Stock Photo; p. 35: PHOTO MEDIA/ClassicStock/Alamy Stock Photo; p. 36: Worldwide Photography/Heritage Image Partnership Ltd/Alamy Stock Photo; p. 37: Pictorial Press Ltd/Alamy Stock Photo; p. 52: Ewing Galloway/Alamy Stock Photo; p. 53: STOCKFOLIO®/Alamy Stock Photo; p. 54: Helmut Meyer zur Capellen/imageBROKER/Alamy Stock Photo; p. 55: L. Banner/ClassicStock/Alamy Stock Photo; p. 57: Slim Plantagenate/Alamy Stock Photo; p. 75 left: Martyn Goddard/getty images; p. 75 right: Bettmann/getty images; p. 127 left: Apic/getty images; p. 127 right: Jean-Louis SWINERS/getty images; p. 153 left: Fairfax Media Archives/getty images; p. 153 right: Paolo KOCH/getty images; p. 194: Hager fotografie/Alamy Stock Photo; p. 195: M&N/Alamy Stock Photo; p. 196: Patti McConville/Alamy Stock Photo; p. 197: Elizabeth Whiting & Associates/Alamy Stock Photo; p. 198: Paul White 1980s Britain/Alamy Stock Photo; p. 199: Jean-Claude Francolon/Gamma-Rapho/getty images); p. 245 left: Keystone-France/getty images; p. 245 middle: Universal History Archive/getty images; p. 245 right: Tony Henshaw/Alamy Stock Photo

Co-Editor's Acknowledgements

Special thanks to
Alan Palmer, Alister Jack, Boston Gielens,
Christian Molz, David Faure, Gregor Wildermann,
Hans Elkerbout, Henk Jan Drenthen, Julia Dauksza,
Leo Petit, Lina Gildenstern, Liza Gielens,
Lopetz Gianfreda, Lou Beeren, Martijn Koch,
Martina Etti, Norbert Hammer, Peter Stut,
René Schoedon, Ruiyao Xu, Simon Hammer,
Sita Permoser, Till Kreutzer, Tobias Sauter.

In memory of
M.J. Gielens
Johann Stoeten
Lilian Van Stekelenburg

Soft Electronics

Iconic Retro Designs from the '60s, '70s, and '80s

This book was conceived, edited, and designed by gestalten.

Edited by Robert Klanten, Elli Stühler, and Rosie Flanagan
Contributing editor: Jaro Gielens

Introduction by Alice Morby
Decade texts by Alice Morby
Brand texts by Daisy Woodward
Product texts by Elli Stühler, in collaboration with Jaro Gielens

Editorial Management by
Lars Pietzschmann

Design, layout, and cover by
Isabelle Emmerich

Photo Editor:
Valentina Marinai

Typeface:
Rotis Sans Serif by Otl Aicher

Product photography by Studio Sucrow
Packaging scans by Jaro Gielens

Printed by Printer Trento s.r.l., Trento, Italy
Made in Europe

Published by gestalten, Berlin 2022
ISBN 978-3-96704-040-1

© Die Gestalten Verlag GmbH & Co. KG, Berlin 2022

All rights reserved. No part of this publication may be reproduced or transmitted in any form or by any means, electronic or mechanical, including photocopy or any storage and retrieval system, without permission in writing from the publisher.

Respect copyrights, encourage creativity!

For more information, and to order books, please visit www.gestalten.com

Bibliographic information published by the Deutsche Nationalbibliothek. The Deutsche Nationalbibliothek lists this publication in the Deutsche Nationalbibliografie; detailed bibliographic data is available online at www.dnb.de

None of the content in this book was published in exchange for payment by commercial parties or designers; gestalten selected all included work based solely on its artistic merit.

This book was printed on paper certified according to the standards of the FSC®.

Jaro Gielens

is a Dutch collector of vintage consumer electronics and electronic games. His collection includes more than 1,200 boxed small household appliances. As an online designer, he has created and designed online ads and websites on a daily basis for the past 25 years. His first book with gestalten, *Electronic Plastic,* explored handheld and tabletop devices from the 1970s and '80s.

www.soft-electronics.com